Introduction to Food Manufacturing Engineering

Tze Loon Neoh • Shuji Adachi • Takeshi Furuta

Introduction to Food Manufacturing Engineering

 Springer

Tze Loon Neoh
Okawara Mfg. Co., Ltd.
Shizuoka, Japan

Takeshi Furuta
Graduate School of Engineering
Tottori University
Tottori, Japan

Shuji Adachi
Graduate School of Agriculture
Kyoto University
Kyoto, Japan

ISBN 978-981-10-9164-3   ISBN 978-981-10-0442-1   (eBook)
DOI 10.1007/978-981-10-0442-1

Printed on acid-free paper

This Springer imprint is published by Springer Nature
The registered company is Springer Science+Business Media Singapore Pte Ltd.

To the memory of my father; to my wife, Patricia, for her unconditional support; and to my family.

T.L.N.

Preface

Food science refers to an integrated science, encompassing comprehensive aspects of the processing of all sorts of bioresources including agricultural, livestock, and marine products into food or food ingredients. Also included are the methods of preparation as well as the effects on the human body resulting from the ingestion of the prepared food products, and in this way food science contributes to the betterment of health and the enrichment of life for mankind. In order to achieve these ultimate goals, food-manufacturing knowledge for processing bioresources into food and food ingredients is therefore indispensable.

The discipline of food manufacturing includes both food processing or food technology, which refers to the study of the objectives and principles of the respective unit processes along the production flow of a food or food ingredient; and also food engineering, which covers the study of a long list of principles common for operations in food-manufacturing processes such as heating and agitation. The former study conforms to actual food-manufacturing flows and thus is readily comprehensible, whereas the latter has high versatility covering the common principles of operations for most food-manufacturing processes. Although many textbooks on food processing are easily accessible, textbooks on food engineering are scarce at the present time, and even those available on the market include a vast array of topics and thus many of them are unsuitable for beginners. Hence, a simple textbook intended for undergraduate students majoring in food-related disciplines other than engineering—for instance, food science, food technology, or nutrition—is highly desirable.

From the standpoint of academics and researchers in the fields of agriculture and engineering, we therefore drew up a plan for this particular book for under-graduate students studying agriculture and engineering. This volume is meant to be an introduction to food engineering in which common principles for food-manufacturing-process operations are taught. Food engineering itself is rather an abstract and hard-to-understand discipline. Therefore, we have endeavored to ini-tially set forth food-manufacturing flows and pay careful attention to quantitatively detailing and explaining the manufacturing operations involved from an engineering point of view. Because the text is considered an "extreme" introductory book to food engineering for beginners, it does not cover all food-engineering-related fields. Rather, it is written for its readership to gain an engineering perspective on the subject and consolidate the foundation of knowledge such that the readers, when necessary, can better understand other more structured and systematically written textbooks on food engineering.

Food engineering deals with a great number of mathematical formulas and numerical values. Because it is difficult for one to acquire such knowledge merely through reading, we have focused on introducing as many example problems and exercises as possible. Furthermore, in many cases graphic representation of data better explains a phenomenon, making it simpler to understand. Graph construction is the most basic but it is seldom covered in lectures at university level, resulting in many students not being able to draw proper graphs. We have illustrated how proper figures are produced and included many sample questions and exercises of which the solutions are obtainable from constructed figures. When it comes to dealing with mathematical formulas and numerical values, practicing problem solving will definitely foster a better understanding, and it is of utmost importance to actually try to put the graphs on paper. Although it has become common for these problem-solving steps to be done on a computer nowadays, one could hardly get a good grasp of the knowledge by just tapping the keyboard and looking at the monitor. On that account, a great deal of thought has been put into selecting the sample questions and exercises such that solving the problems will necessitate sketching and computing with a scientific calculator.

In food engineering, one studies the common principles for process operations, and the knowledge is not limited to food manufacturing but is applicable as well in other areas—for instance, the chemical, pharmaceutical, and environmental industries. Readers are expected to acquire the basic concepts through this textbook and subsequently further deepen the level of understanding via other systematic reference materials.

Shizuoka, Japan   Tze Loon Neoh
Kyoto, Japan   Shuji Adachi
Tottori, Japan   Takeshi Furuta
March 2016

Contents

Chapter 1
Contributions of Food Engineering to Everyday Meals

Abstract The food and food ingredients we consume in our daily meals are mostly produced in great quantities in factories. Although the operations involved in the industrial manufacturing of food share many similarities with normal cooking operations, they also vary in many other aspects. Food processing and food engineering are two disciplines which are concerned with the industrial manufacturing of food. The former studies the respective unit processes along food production flows, while the latter studies principles and analysis approaches common for operations in food-manufacturing processes (unit operations) and thus covers a broad range of applications. In this chapter, we will present a brief overview of unit operations in food engineering. In addition, we will also touch on the points of attention in studying engineering subjects related to food.

Keywords Food engineering • Food manufacturing • Cooking operations • Unit operations • Food science

1.1 Diets and Food Ingredients

Imagine one day you have a glass of cow milk to go with a slice of margarine-spread toast for breakfast; at noon, you have Italian spaghetti and green salad dressed with mayonnaise for lunch and then an after-meal drink of instant coffee; and then for dinner, you have rice as your staple and sashimi, tempura, miso soup with tofu, and pickles as accompanying dishes. Meanwhile you also enjoy a glass of shochu on the rocks with dinner (Fig. 1.1). Table 1.1 summarizes the foods and main food ingredients used for that day's meals. Other than those like vegetables that could be consumed directly from the farm or after some home cooking, there are many factory-made products such as bread and margarine, that are rarely prepared at home but instead are mostly purchased from food markets. Among all the foods and food ingredients illustrated in Fig. 1.1, the majority of the items, namely, bread, margarine, milk, spaghetti, ketchup, mayonnaise, instant coffee, miso, and shochu, are ready-made products industrially processed and produced for consumers' convenience. So now, you may start wondering as to how those foods and food ingredients are prepared or perhaps whether the wheat flours used to produce various food products as in bread, spaghetti, and tempura are of the same kind.

© Springer Science+Business Media Singapore 2016   1
T.L. Neoh et al., *Introduction to Food Manufacturing Engineering*,
DOI 10.1007/978-981-10-0442-1_1

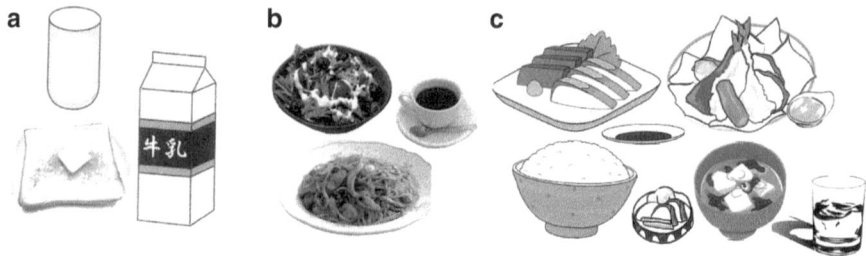

Fig. 1.1 Foods for (**a**) breakfast, (**b**) lunch, and (**c**) dinner

Table 1.1 Main ingredients of typical foods consumed in meals of a day

	Foods	Main ingredients
Breakfast	Bread loaf	Wheat flour, baker yeast
	Margarine	Oil and fat, salt, emulsifier
	Processed milk	
Lunch	Spaghetti	Wheat flour, ketchup
	Salad	
	Mayonnaise	Vegetable oil, egg, vinegar
	Coffee	Coffee
Dinner	Rice	
	Miso soup	Soy bean, salt
	Sashimi	
	Tempura	Wheat flour
	Shochu	Sweet potato, wheat, etc.

1.2 Common Operations in Preparation of Food/Food Ingredients

The foods and food ingredients listed in Table 1.1, e.g., bread, dry spaghetti, instant coffee, and milk are basically produced as described below. A fermented sponge dough consisting of well-combined wheat flour, baker yeast, and water is kneaded with ingredients such as wheat flour, sugar, and water, before the final dough is being nicely shaped, proofed, and then baked to prepare bread (Fig. 1.2). Meanwhile spaghetti is prepared by drying fresh cylindrical pasta in hot air (Fig. 1.3). Pasta doughs are also made of wheat flours but of different kinds from those used in bread making and are prepared by kneading the wheat flours with water. The doughs are subsequently extruded through molds with holes commonly known as dies to give them the cylindrical shape. As for instant coffee, the coffee beans are first roasted and then ground before being subjected to hot water extraction. The extract then goes through a concentration process and would finally be dehydrated either by spray drying or freeze drying (Fig. 1.4). The drying method determines the morphology of the final product with the former producing fine free-flowing particles while the latter creating coarse rough ones.

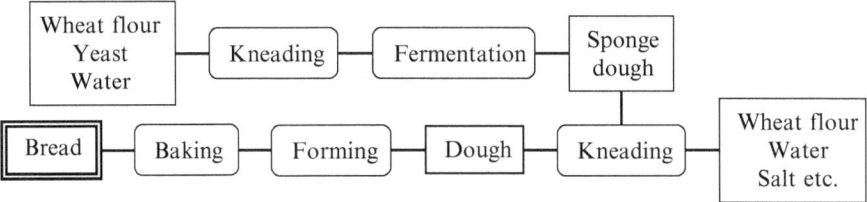

Fig. 1.2 Bread making process

Fig. 1.3 Manufacturing process of spaghetti

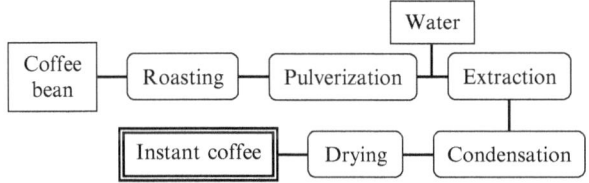

Fig. 1.4 Manufacturing process of instant coffee

Fig. 1.5 Manufacturing
process of processed milk

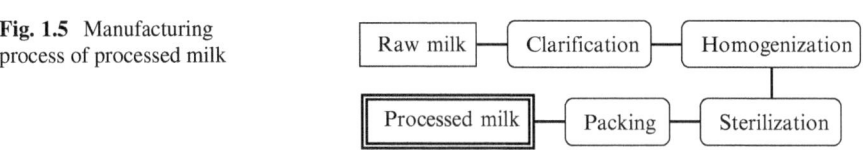

Cow milk in the market is processed from fresh cow milk which is expressed from milk cows' udders as it is without incorporation of any additives. However, the milk fat globules in fresh milk tend to separate out and rise to the surface upon standing during transportation and storage due to the facts that the fat globules are too large and diverse in size to stay suspended within the aqueous environment. The fat globules are hence broken up (homogenized) to smaller sizes such that they remain suspended evenly in the milk to avoid creaming at the surface. Homogenized milk further goes through some heating treatment (pasteurization) in order to kill any harmful bacteria for better shelf life (Fig. 1.5).

Among the above-described food processes, some are common operations involved in the preparation of various foods. For instance, kneading is a common operation for the preparation of both bread and spaghetti, while emulsification is employed in the production of both margarine and milk. Likewise, spaghetti and instant coffee both require a drying process to arrive at their final dry forms. The main processes involved in the production of the foods and food ingredients listed in

Table 1.2 Processing operations involved in the manufacturing processes of main ingredients of typical foods consumed every day

	Foods	Processing operations in the manufacturing processes						
		Heating/sterilization	Concentration	Distillation	Extraction	Kneading/emulsification	Drying	Fermentation
Breakfast	Bread loaf	○						
	Margarine					○		
	Processed milk	○				○		
Lunch	Spaghetti					○	○	
	Mayonnaise					○		
	Coffee		○		○		○	
Dinner	Rice	○						
	Miso soup	○						
	Tempura	○						
	Shochu			○				○

Table 1.1 are summarized in Table 1.2. The enumerated food processing operations include typical everyday operations such as heating as well as those not so common on household level like emulsification which is necessary for the preparation of margarine, mayonnaise, and milk. Emulsification is an operation by which two immiscible liquid phases (aqueous and oil phases) are mixed in order to suspend evenly and stabilize one of the phases (for instance, the oil phase in milk) within the other (the aqueous phase).

Food-manufacturing processes are often made up of combinations of several of these operations such as heating, kneading, and drying. *Food manufacturing* carries the meaning of producing food on a large scale as opposed to similar activities on household level. Individual operations that comprise a food-manufacturing process are regarded as *unit operations*. These unit operations often involve heat and mass transfers, the basics of which are comprehensively detailed in an academic discipline called *transport phenomena*. Besides, *reaction engineering* is a discipline in which reaction-accompanied phenomena are studied quantitatively. Transport phenomena and reaction engineering form along with physical chemistry the foundations of many unit operations. *Food processing* refers to the comprehensive study pertaining to food manufacturing in every single processing stage of a food, like those mentioned earlier in this chapter such as bread and spaghetti, including the principles of the constituent unit operations along the operation flow. Whereas in the field of *food engineering*, one would study the principles and analysis approaches of unit operations that make up a food-manufacturing process, which are also

Table 1.3 Cooking operations and the corresponding unit operations in food engineering

Cooking operations	Processing operations in manufacturing processes												
	Sterilization	Heat transfer	Freezing/thawing	Concentration	Distillation	Extraction	Fluidization	Agitation/emulsification	Pulverization	Filtration	Adsorption/cleaning	Drying	Reaction/fermentation
Stewing		○											
Steaming		○											
Grilling		○											
Stir-frying		○											
Roasting		○											
Boiling		○											
Straining										○			
Making soup stock						○							
Microwave heating		○											
Washing/immersion													
Kneading								○					
Grating/smashing									○				
Sousing/firming/deodorizing						○							
Aging													
Pickling													
Gelling													
Emulsifying								○					
Frothing								○					
Cutting									○				
Mixing								○					
Dressing								○					
Cooling		○											
Freezing			○										
Seasoning													

highly versatile in terms of application in various other food products. For example, the drying operations in the preparation of dry spaghetti and instant coffee are considered a same operation in terms of reduction of moisture content, regardless of the food product in question. In actual food manufacturing, the knowledge in both food processing and food engineering are equally important and necessary as the former explains the principles of the operations employed in the processing of a particular food and the reasons why they are needed and the latter interprets not only the principles but the analysis approaches as well of the operations.

  Cooking operations for household and group catering and their corresponding unit operations in food engineering are shown in Table 1.3. Cooking often involves heating (heat transfer) and mixing (agitation/emulsification) of food. Operations like pickling and seasoning, which allow flavoring materials to soak into food are operations based on diffusion which form the foundation of a number of unit operations in food engineering. On the other hand, food engineering encompasses the manufacturing of food ingredients as well, for instance, wheat flours, sugars, and vegetable oil, and thus covers also those operations that are hardly carried out in cooking such as concentration, distillation, fluidization, membrane separation, drying, reaction, and fermentation. Furthermore, group catering basically involves cooking operations similar to those in household cooking except for the fact that it utilizes bigger-sized utensils solely due to comparatively larger amount of food prepared. Nevertheless, as mass production of food in an efficient manner is necessary in the food industry, not only in terms of scale but the operations themselves as well often differ from ordinary cooking.

1.3   Primary Unit Operations

In cooking, most of the operations like stewing, steaming, and cooling, involve the application and removal of heat. In food manufacturing too, there are many operations that are associated with inflow and outflow of heat such as thermal sterilization, thermal concentration, freezing, and thawing. These operations are so-called *heat transfer operations* because they are always accompanied by transmission of heat.

Sterilization refers to the operation to eliminate microorganisms in food in order to extend shelf life. Sterilization of food may basically be categorized into two major types: (i) thermal sterilization that employs heat to kill pathogenic microorganisms and (ii) cold sterilization (nonthermal sterilization) that leverages chemicals and radiations. However, thermal sterilization remains the mainstream operation in food processing. Canned food and retort pouch are some example food thermally sterilized for better storage stability. Besides, the inhibition of growth and proliferation of bacteria without destroying them is named *bacteriostasis*.

The operation to thicken a liquid food by lowering its water content is called *concentration*. By reducing the volume of a liquid, concentration allows the downsizing of container and also the cutback on transportation cost. Moreover, higher concentrations of liquid also retard the growth of microorganisms as the osmotic pressure increases with concentration, therefore prolonging the shelf life. There are several ways to concentrate a liquid: (i) evaporative concentration that evaporates water by heating, (ii) freeze concentration that crystallizes water out of the liquid phase by cooling, and (iii) membrane concentration that utilizes reverse osmotic membranes to remove water. The concentration of fruit juices are often performed by evaporative concentration.

The operation to transform liquid food to solid form by removal of heat to enable long-term storage is called *freezing*, whereas the reverse operation to defrost frozen food for further processing or consumption is known as *thawing*. Frozen food which is gaining popularity nowadays would make a good example of the food products, the preparation of which requires those two operations.

The vapor produced when heating up an ethanol solution contains higher proportions of ethanol compared to the solution itself, and therefore a solution of higher alcohol (ethanol) concentration than the mother liquid may be collected by condensing the vapor. The separation method that takes advantage of the difference in volatility of liquids is called *distillation* which is a method employed in the preparation of distilled liquors such as shochu and whisky.

Solid-liquid extraction is a usual operation in food manufacturing. An everyday routine of coffee making is a good example of solid-liquid extraction. When hot water is poured into coffee nibs, the soluble components inside the nibs (solid phase) would migrate into the hot water (liquid phase). In addition, migration can also occur from a liquid phase in which the components are initially present to another immiscible liquid phase. In this case, the operation is called *liquid-liquid extraction*.

Tropical fishes kept in an aquarium as pets are supplied with oxygen by bubbling air into the water in order to promote dissolution of oxygen. The operation to dissolve the components in a gaseous medium into a liquid medium is known as *gas absorption*. Cultivation of aerobic microorganisms like baker yeast, hydrogenation of edible oils, carbonation of beverages, etc., involve this kind of operation. The reverse operation of gas absorption in which the components in a liquid phase are released into the gaseous phase is called *stripping*.

The water filter with a filter cartridge filled with activated carbon installed to the tip of a faucet would eliminate certain impurity components in the tap water by *adsorption*. Trapping the components that are present in a liquid and a gas by a solid (adsorbent material) are known as liquid phase adsorption and gas phase adsorption, respectively. In food processing, liquid phase adsorption by activated carbon is employed for decolorization of sugar concentrate.

Operations that involve phase transition and migration of substances from a phase to another are collectively regarded as *diffusional unit operations*, whereas those that do not involve the two are generically known as *mechanical unit operations*, for instance, agitation, emulsification, pulverization, and filtration.

The unit operation that involves the combination of dissimilar ingredients into one even mass is generally termed *mixing*. The operation of stirring a tank of liquid such that the ingredients and temperature are uniform is called *stirring*, while in the case that the liquid is very viscous, the operation of vigorous stirring would then be regarded as *agitation*. Furthermore, the process of blending of materials in powder form like wheat flour with a small quantity of water into homogenous dough is called a *kneading* operation which is required for the preparation of bread or spaghetti as mentioned earlier in this chapter. The process that mixes two or more immiscible liquids such that droplets of one of the liquids are suspended evenly within the matrices of the other(s) is referred to as an *emulsification* operation. Mayonnaise and margarine are examples of food products of which the manufacturing processes include emulsification.

Wheat flours are obtained by grinding down wheat grains to fine particles, which is known as *pulverization*. The pulverized particles are usually irregular in respect of size (particle diameter) and would be further separated into certain size groups by *classification*.

Filtration refers to the operation to strain liquid components off a slurry to separate the dispersed solid components. For instance, this operation is done in beer breweries to remove brewer's yeast from fermentation liquid. Besides, the operation to squeeze out the liquid portions of the fermentation broths of alcoholic beverages and shoyu (soy sauce) is known as *pressing* or *expression*.

Drying is an operation which eliminates moisture. The production of instant coffee involves drying of coffee extract into solid. Heat of vaporization which could be supplied by media such as hot air, superheated steam, and inert gases, is required for water to undergo the change from the liquid phase to the gaseous phase. Although drying is an extensively used common operation in the food industry, it is also a very complicated phenomenon because it involves simultaneous transfers of mass (water) and heat.

Some food processing involves transformation of substances into some other things by chemical reaction, which is generically known as a reaction operation. The engineering design and handling of reactors are referred to as *reaction engineering*. The reactors that employ biocatalysts such as enzymes and microorganisms are known as *bioreactors*. The fructose-glucose syrup (high-fructose corn syrup) in drinks, confectioneries, and such is produced by the use of immobilized enzymes (enzymes fixed on water-insoluble carrier) in bioreactors. The food processing that harnesses the power of microorganisms as exemplified by the production of Japanese sake, beer, miso (fermented bean paste), shoyu, and etc., is called *fermentation*. In such operation, the reactors are named fermenters. On the contrary, the reactions, also facilitated by microorganisms, that lead to the production of undesirable substances causing a food product to perish are called *decay* or *putrefaction*.

In food processing, the flow conditions of fluids (inclusive of both gases and liquids) inside conduits and containers exercise strong influences on the heat and mass transfers. In the case of flowing fluids inside conduits, friction against the conduit wall causes a loss of energy and a decrease in pressure, a phenomenon known as *pressure drop*. These items are all covered under *fluid flow*. The surface of a konjac slab dents when you press it with a finger and snaps back upon removal of the applied pressure. If you apply mild pressure over the top surface of the slab with the palm of your hand and try to slide it in a horizontal direction, it deforms obliquely. If you try to stir a cup of honey with a teaspoon, the honey hardly moves let alone swirl except in the vicinity of the teaspoon. The farther it is away from the teaspoon, the stiller it stands. Food processing would also necessitate the understanding of how food materials deform with applied forces. These mechanical behaviors are detailed in a discipline called *rheology* with fluid flow also being part of it.

Despite the fact that food products are usually stored in refrigerators, freezers, etc., at low temperatures in low humidity conditions, most food products, though at relatively slower rates, still do degrade in terms of quality during storage. On account of that, the comprehension of factors that affect the quality deterioration of food products is required for food manufacturing.

1.4 Facing the Challenge

Preparation of food does not require special education. You all supposedly have already had some, if not plenty of, cooking experience before taking up the courses in food engineering and food processing. Does that thus make food manufacturing a piece of cake?

The quality of raw materials for food and food ingredients may vary with region as well as season and year of harvest. Nonetheless, there is no excuse for food products to evade certain defined standards meant to ensure food quality. Besides, the raw materials and their products are the so-called living matters and therefore are prone to decay and discoloration causing not only deterioration of quality but sometimes, in a worst-case situation, food poisoning as well. Because food or food ingredients usually consist not of single substances but instead of multiple substances, the occurrence of physical, chemical, or biological interactions between the constituents is not uncommon. These multiple substances are often present in unequal proportions with certain trace components frequently playing important roles in giving the foods and food ingredients their characteristic aromas and flavors. Moreover, it is of pivotal importance for a food product to be definitely safe for large consumption (within sensible range) over lengthy periods of time.

Let us compare the properties of food with medicines that are likewise administered orally. Excluding the excipients, pharmaceutical products are basically made up of single substances, and the raw materials are not necessarily of naturally occurring origins. And for many of them, quality deterioration occurs on longer time scales relative to food products. Sometimes some side effects and even high prices are tolerable depending on disease condition. On the other hand, food products are supposed to be tasty though the criteria for palatability may differ substantially depending on country, region, generation, etc. In addition to its primary functions to sustain life by providing nourishment and its secondary functions to satiate one's craving for taste, liking for color, etc., food has recently been elucidated to serve tertiary functions in aiding regulation systems in living organisms such as biological defense, regulation of biorhythm, aging control, prevention of diseases, etc. However, these effects brought about by ingestion of food are generally sluggish, and scarcely are food able to produce quick responses like other pharmaceutical products do.

Although there are great differences between food and pharmaceutical products, neither of them are superior to the other. Medications that demonstrate specific effectiveness in the treatment of severe diseases may make unfathomable evangel for patients. Nevertheless, the absolute number of people who benefit from pharmaceutical products in this manner is usually small. On the other hand, every single one of us has been receiving the benefits of nourishment (primary functions), deliciousness (secondary functions), and the modest biological regulation functions (tertiary functions) from the food we consume on a daily basis. Had it been possible to measure happiness by the product of intensity and quantity, there might though be differences in intensity and quantity between pharmaceutical products and food, their products would have been much the same as both of them are contributing to the well-being and life enrichment of mankind.

Food science refers to the integrated science that covers various aspects of food including broad academic knowledge and methodology, which means that a food science student would acquire a nodding acquaintance with both basic sciences like chemistry, physics, biology, biochemistry, mathematics, and microbiology, and applied sciences such as nutrition, food chemistry, and hygiene. Food engineering, which is a fundamental science of food manufacturing, is also one of the applied sciences to be learned in this discipline.

Exercise

1.1 Pick up a dish of your last dinner and find out the manufacturing processes of the ingredients.
1.2 What are the other food ingredients which are also manufactured using the unit operations involved in the manufacturing processes you mentioned above in Exercise 1.1?
1.3 Find out the varieties and applications of rice flours produced by pulverization of different varieties of rice grains.

Chapter 2
Bookkeeping of a Process

Abstract Not only numerical numbers but also units are essential for quantitative expression of physical quantities. We first give a general account of the internationally standardized system of units called the International System of Units (SI). Food processing consists of processes which have been rationally designed and optimally operated based on mass and energy balances. The chapter also discusses how mass balances for steady-state and unsteady-state operations are set up. We then review some logarithmic identities as they are useful for adequately describing many phenomena occurring in food processing operations. Further, we address issues about how to draw graphs as a dominant tool for quantitatively expressing and aiding in the understanding of various phenomena.

Keywords Dimension • Unit • SI • Mass balance • Steady state • Unsteady state • Key component • Flow sheet • Differential equation • Index number • Logarithmic number • Graph drawing

2.1 Units

Length, time, mass, temperature, etc., are basic concepts called *dimensions* for expressing physical quantities. Further, these physical quantities are represented quantitatively by means of *units*. Some physical quantities, e.g., length and time, are expressed by the *base units* ([m] and [s], respectively) (Table 2.1), while others, e.g., area and velocity, can be expressed as combinations of the base units ($[m] \times [m] = [m^2]$ and $[m] \div [s] = [m/s]$, respectively), which are known as the *derived units*.

A number of systems of units have been in use in the past depending on the set of physical quantities chosen as the base units, causing inconvenience in many ways. The *International System of Units* (Le Systeme International d'Unites (French); abbreviated as *SI*), a reasonably systemized absolute system of units based on the fundamental base units of length, mass, and time, makes the most widely used system of units nowadays.

© Springer Science+Business Media Singapore 2016
T.L. Neoh et al., *Introduction to Food Manufacturing Engineering*,
DOI 10.1007/978-981-10-0442-1_2

Table 2.1 SI base units

Physical quantity	Name	Symbol
Length	Meter/meter	m
Mass	Kilogram	kg
Time	Second	s
Electric current	Ampere	A
Thermodynamic temperature	Kelvin	K
Amount of substance	Mole	mol
Luminous intensity	Candela	cd

Table 2.2 SI derived units with special names

Physical quantity	Name	Symbol	Definition based on SI base and derived units
Force	Newton	N	$m \cdot kg/s^2$
Pressure/stress	Pascal	Pa	$kg/(m \cdot s^2) = N/m^2$
Energy (work, heat)	Joule	J	$m^2 \cdot kg/s^2 = N \cdot m$
Power	Watt	W	$m^2 \cdot kg/s^3 = J/s$
Quantity of electricity (electric charge)	Coulomb	C	$s \cdot A$
Voltage (electrical potential difference)	Volt	V	$m^2 \cdot kg/(s^3 \cdot A) = J/(A \cdot s) = J/C$
Electrical resistance	Ohm	Ω	$m^2 \cdot kg/(s^3 \cdot A^2) = V/A$
Electrical conductance	Siemens	S	$s^3 \cdot A^2/(m^2 \cdot kg) = A/V = \Omega^{-1}$
Electrical capacitance	Farad	F	$s^4 \cdot A^2/(m^2 \cdot kg) = A \cdot s/V$
Frequency	Hertz	Hz	s^{-1}
Celsius temperature	Degree celsius	°C	$t\,[°C] = T\,[K]\ 273.15$ (temperature difference) $1\,°C = 1\,K$

Table 2.1 shows the seven SI base units from which other SI units are derived. Molecular mass is intrinsically a dimensionless quantity, but it is presented in the units of [kg/mol] in equations for the sake of dimensional soundness according to the definition of amount of substance. The SI unit for amount of substance is the mole [mol]. The physical quantity described by the units of [kg/mol] is regarded as molar mass.

Table 2.2 summarizes some of the common SI derived units out of the 17 named ones defined from the seven SI base units.

In the SI, a physical quantity is presented only in one type of unit(s). Therefore, the conversion factor between any physical quantities within the system is always 1. Nonetheless, quantities presented in the SI base units and the SI derived units may not necessarily be convenient in practice. For instance, the length of the railroad from Kyoto to Tokyo is approximately 513,600 m, $513{,}600\ m = 513.6 \times 10^3\ m = 513.6\ km$; it is simply handy to add the *prefix* k (kilo) that denotes a multiple of a thousand to the unit of [m] (meter). Application of prefixes (Table 2.3) is allowed by the SI.

Table 2.3 SI prefixes

Factor	10^{-1}	10^{-2}	10^{-3}	10^{-6}	10^{-9}	10^{-12}
Name	Deci	Centi	Milli	Micro	Nano	Pico
Symbol	d	c	m	μ	n	p
Factor	10^{1}	10^{2}	10^{3}	10^{6}	10^{9}	10^{12}
Name	Deca	Hecto	Kilo	Mega	Giga	Tera
Symbol	da	h	k	M	G	T

2.2 Mass Balance

2.2.1 Mass Balance: Meaning and Significance

Various kinds of substances contained in foods undergo chemical or physical changes during processing which also involves various types of energy. These substances and types of energy may transform into different forms, but they can never be created or destroyed as stated in the laws of conservation of mass and energy. It is of utmost importance to master this knowledge for the design and operation of processes and equipment. Here, we will further discuss *mass balance* based on the *law of conservation of mass* by explaining mainly through example questions.

To determine a mass balance, a target scope (known as a "system") has to be first selected before we start to figure out the balance between the substances that go into and come out of the system. In the case where chemical reactions take place, the generation and loss of materials shall be taken into account. In household bookkeeping, the amount of savings can be obtained by subtracting the expenditures from the incomes. Likewise, the basic equation for mass balance is given by

$$\text{(Accumulation)} = \text{(Input)} - \text{(Output)} + \text{(Generation)} - \text{(Loss)} \qquad (2.1)$$

When mass balance per unit time is considered, the equation can be expressed in terms of rate of each component as

$$\text{(Accumulation rate)} = \text{(Input rate)} - \text{(Output rate)} + \text{(Generation rate)} - \text{(Loss rate)} \qquad (2.2)$$

A mass balance is known as being in *steady state* when the left side of both Eqs. (2.1) and (2.2) are equal to 0, a condition in which the accumulation neither increases nor decreases. In addition, as for mass balances with chemical reactions, computation based on the amount of substance in the SI unit of mole may often turn out to be relatively simpler.

Why ever are we interested in studying mass and energy balances? You may understand better as we carry on. For instance, mass and energy balances may serve a useful purpose for understanding the consistency of equipment and plants. Let us say we are measuring the input and output quantities of a process; we can check out the accuracy of the measurements based on the mass balance relation of the process. If the components do not balance out each other within the tolerable accuracy, it indicates that either there might be errors in the measurements or leakage and/or accumulation within the system. Furthermore, the input and output of many processes are not directly measurable, and in this kind of processes, the unknown quantities may be estimated on the basis of the mass balance relations.

2.2.2   Calculation Steps of Mass Balance

Balances are generally calculated according to the following steps. Firstly, (1) draw a simple diagram (*flow sheet*) of the process. Next, (2) fill out the known quantities on the flow sheet, and insert variable symbols for unknown quantities. Then, (3) write down the chemical reaction equation if chemical reactions take place during the process. (4) Select an appropriate basis for calculation and put it down on the flow sheet. At this point, the assigned basis may still not be the most convenient one. Lastly, (5) apply Eq. (2.1) or Eq. (2.2) and write up the mass balance equations for all substances and also for each constituent. A high number of unknown quantities may complicate the calculation. On the other hand, there may exist substances which, even after having gone through the process, are unaltered, and these substances are able to provide clues which often simplify the calculation of a mass balance.

Let us go through some examples below so that you get a better grasp about mass balances.

Example 2.1   Seawater contains roughly 3.5 % (w/w) of salt. How many kilograms of anhydrous salt can one get out of 1.0 kg of seawater by drying up the water in it? And how many kilograms of water are there that need to be evaporated?

Solution   Setting 1.0 kg of seawater as the calculation basis, let the amount of water that needs to be evaporated and the amount of salt obtainable after evaporation be m_W [kg] and m_S [kg], respectively. The flow sheet can be constructed as depicted in Fig. 2.1. Considering the dashed line-framed system, the system has an input of 1.0 kg of seawater and two respective outputs of water, m_W [kg], and anhydrous salt, m_S [kg]. In addition, there are no other substances being generated or vanished. Hence, the mass balance equation for all input and output substances of the system can be written in terms of the known and unknown quantities as

$$0 = 1.0 - (m_W + m_S) \tag{2.3}$$

Next, because the salt contained in the seawater is discharged from the system without undergoing any reactions, it may provide a clue to the calculation. The concentration of salt in the seawater is presented in the units of % (w/w), known as *percentage concentration by weight*, which is a representation in percentage of the mass of salt [kg] in a unit mass (e.g., 1 kg) of seawater. Refer to Appendix A1 for

Fig. 2.1 Flow sheet of an evaporation process of seawater

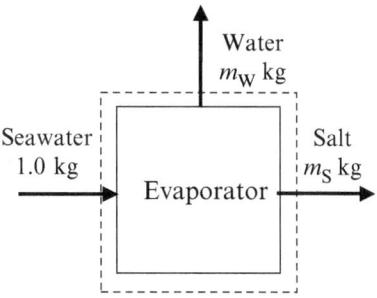

various expressions of the physical quantity of concentration. The mass of seawater consists of the masses of water and salt and therefore % (w/w) can be described as

$$\% \ (\mathrm{w/w}) = \frac{\mathrm{kg\text{-}salt}}{\mathrm{kg\text{-}seawater}} \times 100 = \frac{\mathrm{kg\text{-}salt}}{\mathrm{kg\text{-}water} + \mathrm{kg\text{-}salt}} \times 100 \qquad (2.4)$$

Hence, the mass balance equation for the salt is given by Eq. (2.5):

$$(1.0)(0.035) = m_\mathrm{S} \qquad (2.5)$$

By Eqs. (2.5) and (2.3), m_S and m_W are determined to be 0.035 kg and 0.965 kg, respectively. ◢

Example 2.2 Spaghetti is produced from wheat middlings called durum semolina kneaded with water (refer to Fig. 1.3). A fresh pasta dough was prepared from 700 g of durum semolina with 14 % moisture on wet basis kneaded with 330 g of water. The dough was then extruded through a die with round holes into spaghetti. If the fresh spaghetti is to be subjected to a drying process until it attains a dry basis moisture content of 11 %, how many grams would the dried spaghetti weigh?

Solution The amount of moisture contained in foods is called moisture content. Because moisture content is a term with ambiguous definition, it often causes some confusion in the research and industrial fields. For instance, when 100 g of durum semolina is dried thoroughly, the sample weight decreases to 86 g. In this case, we know that the durum semolina contains $100 - 86 = 14$ g of water. The moisture content of this sample of durum semolina can be presented in two different manners. One is the *wet basis moisture content* given by the weight ratio of water to the original moisture-containing durum semolina (wet material). The wet basis moisture content is equal to $14/100 = 0.14$ (or 14 %) in the units of [g-water/g-wet material]. The other way is by the *dry basis moisture content* defined as the weight ratio of water to the bone-dry durum semolina (dry material). The dry basis moisture content of the durum semolina sample is $14/86 = 0.16$ (16 %) in the units of [g-water/g-dry material]. Sometimes "dry matter" or "dry solid" is used instead of "dry material," and they are represented by their respective acronyms of d.m. and d.s. in the units as [g-water/g-d.m.] or [g-water/g-d.s.]. In the field of food engineering, the term

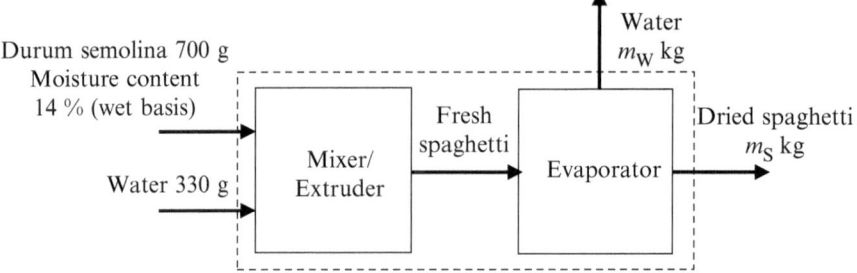

Fig. 2.2 Flow sheet of the manufacturing process of dried spaghetti

"moisture content" alone usually refers to dry basis moisture content. Care should
be taken because sometimes, but not often, it is confused with wet basis moisture
content.

The values of moisture content vary by definition, and the higher moisture a
food contains, the greater the difference. For example, 100 g of spaghetti cooked
until ready to eat contains roughly 70 g of water. The wet basis moisture content
of this spaghetti is $70/100 = 0.70$ (70 %), while the dry basis moisture content is
$70/(100 - 70) = 2.33$ (233 %). There is such a huge difference in terms of figures.
Wet basis moisture content can never exceed 100 % by definition, whereas it is not
rare for dry basis moisture content to go beyond 100 %.

We will consider this example question paying attention to the aforementioned
aspect. The manufacturing process of dried spaghetti consists of the operation of
kneading durum semolina with water and the operation of drying fresh spaghetti. Let
us draw a flow sheet of the process (Fig. 2.2) and fill out the information available.
We set 700 g of durum semolina with a wet basis moisture content of 14 % as the
calculation basis. Bone-dry durum semolina may serve as a clue substance because it
is unaltered through the process. We let the outputs of dried spaghetti and water from
the drying operation be m_S [g] and m_W [g], respectively. First of all, we will consider
the mass balance of all substances. Because the kneading and drying operations
neither bring about generation of substances nor incur any losses, the amounts of
input and output substances within the dashed line-framed system are equal, and
hence the following equation holds

$$700 + 330 = m_S + m_W \tag{2.6}$$

Next, let us consider the mass balance of the clue substance. Durum semolina
as the raw ingredient contains 14 % moisture on wet basis, and thus 700 g of
durum semolina consists of $(700)(1 - 0.14)$ g bone-dry durum semolina. On the
other hand, the dried spaghetti contains 11 % moisture on dry basis. The wet basis
moisture content, w, can be computed from the dry basis moisture content, X, by the
equation below:

$$w = X/(1 + X) \tag{2.7}$$

The moisture content of 11 % on dry basis is equivalent to $0.11/(1 + 0.11) = 0.099$ or 9.9 % on wet basis. Hence, m_S [g] of dried spaghetti contains $m_S (1 - 0.099)$ [g] of dry material which is equivalent to the dry material input to the system:

$$(700)(1 - 0.14) = m_S (1 - 0.099) \tag{2.8}$$

m_S is thus determined to be 668 g by Eq. (2.8) and by plugging this value into Eq. (2.6), m_W is calculated to be 362 g. ◢

For systems wherein substances are transformed into other forms via chemical reactions, it may be simpler to look at the mass balances in terms of the amount of substance (in the unit of mole).

Example 2.3 Three hundred grams of warm water was added to 700 g of anhydrous sucrose, and then a small amount of invertase (enzyme) was added to catalyze the hydrolysis of sucrose. Sixty-five percent of the sucrose was hydrolyzed into glucose and fructose. The sugar resulted from invertase-catalyzed hydrolysis of sucrose is called inverted sugar. What is the proportion (weight ratio) of sucrose, glucose, fructose, and water in the resultant inverted sugar syrup? Let the respective molar masses of sucrose, glucose, fructose, and water be 342, 180, 180, and 18 g/mol.

Solution Hydrolysis of sucrose can be described by

$$\text{Sucrose} + \text{water} \rightarrow \text{glucose} + \text{fructose} \tag{2.9}$$

In terms of the amount of substance, 1 mol of water is required to hydrolyze 1 mol of sucrose and results in 1 mol each of glucose and fructose. Seven hundred grams of sucrose is equal to $700/342 = 2.05$ mol of sucrose. We know that 65 % of it underwent hydrolysis; thus, the amount of water molecules consumed and the amounts of glucose and fructose molecules generated are $(2.05)(0.65) = 1.33$ mol. And the amount of residual sucrose molecules would be $(2.05)(1 - 0.65) = 0.72$ mol (or $2.05 - 1.33 = 0.72$ mol). Meanwhile, 300 g of water is equivalent to $300/18 = 16.67$ mol of water molecules. During hydrolysis of sucrose, 1.33 mol of water is consumed, resulting in a residual of $16.67 - 1.33 = 15.34$ mol. The composition of the sugar syrup in terms of the amount of substance resulted from the hydrolysis reaction of 65 % of sucrose is summarized in Table 2.4. The weight of each of the constituent substances can be determined by multiplying the amount of substance by the corresponding molar mass, and the weight ratio of each substance can subsequently be obtained as listed in the rightmost column of Table 2.4.

The mixture of 700 g of sucrose and 300 g of water weighs 1000 g, and the sum of weights of all the constituent substances after hydrolysis reaction must also be 1000 g. However, you may have realized that the sum of weights of all the substances is 1001.1 g. The difference of 1.1 g is not because of the fact that something was generated from the hydrolysis reaction but is instead caused by rounding errors during the calculation process. ◢

Table 2.4 Composition of a sugar syrup

Composition	Amount [mol]	Weight [g]	Weight ratio
Sucrose	0.72	246.2	0.246
Water	15.34	276.1	0.276
Glucose	1.33	239.4	0.239
Fructose	1.33	239.4	0.239
Total	18.72	1001.1	1.000

2.2.3 Mass Balance of Unsteady-State Process

A process is known to be in an *unsteady state* if the left side of Eq. (2.1) or Eq. (2.2) of the process is not 0, meaning that substances within the system either accumulate or diminish. Mass balances of processes in unsteady-state regimes are often represented using differential equations.

Example 2.4 One kilogram of seawater (salt concentration: 3.5 % (w/w)) was heated using an electric heater (output: 1 kW) in a pot for concentrating the seawater through evaporation of water. After the seawater was brought to a boil, water was evaporated at a rate of 16 g/min. Establish the relationship between the salt concentration, w_S [g-salt/g-solution], and the evaporation time, t [min]. And determine also the efficiency of the electric heater. Let the heat of evaporation (heat of vaporization) of water be 2.25 kJ/g.

Solution Consider the pot containing seawater as a system; the overall mass balance can be derived from Eq. (2.2) as follows because there is no input substance while the output is comprised of only the evaporated water:

$$\text{(Accumulation rate)} = - \text{(Output rate)} \tag{2.10}$$

Let the weight of seawater in the pot be W [g], and the change rate (accumulation rate) of it can be expressed as dW/dt [g/min]. Hence,

$$\frac{dW}{dt} = -16 \tag{2.11}$$

The initial weight of seawater is 1000 g (this is the *initial condition* for Eq. (2.11)), and thus the relationship between the weight of seawater, W, and time, t, can be described by

$$W = 1000 - 16t \tag{2.12}$$

Equation (2.12) can be determined from Eq. (2.11) as explained below. d is the derivative symbol which denotes an infinitely small change in a variable. dW and dt represent infinitesimal changes in the weight of seawater [g] and time [min], respectively. The ratio between the two, dW/dt, denotes the weight change with respect to infinitesimal change in time, which is the rate of weight change of seawater [g/min]. Moving dt (the term denotes an infinitesimal change in t, so it is treated as an entity instead of $d \times t$) in Eq. (2.11) to the right side gives

$$dW = -16dt \tag{2.13}$$

The variables W and t now occur, respectively, on the left and right sides of the equation. This manipulation is called separation of variables. A differential equation is called a *variables separable differential equation* if the variables can be separated with one variable on one side of the equation and the other on the other side via the described manipulation like Eq. (2.11) (see Appendix A3). Note that the term dW on the left side of Eq. (2.13) is multiplied by a constant of 1. Integrating Eq. (2.13) gives

$$\int (1)dW = \int (-16)\, dt + c \tag{2.14a}$$

$$W = -16t + c \tag{2.14b}$$

where c is the constant of integration. Equation (2.14b) is the generalized solution for Eq. (2.14a), but in the field of engineering, a solution that satisfies certain conditions (known as a *particular solution* in which the arbitrary constant of integration is given a particular value) would be necessary. The aforementioned initial condition of seawater weighing 1000 g is a condition for Eq. (2.11), wherein $W = 1000$ when $t = 0$. Substituting the values of the initial condition into Eq. (2.14b) gives $c = 1000$, and thus Eq. (2.12) is obtained.

As described in the above paragraph, the constant of integration, c, in Eq. (2.14b) is obtained by indefinite integration of both sides of Eq. (2.13) and successive substitution of the initial condition, thus yielding Eq. (2.12). When we integrate Eq. (2.13), if we use the initial condition of $W = 1000$ when $t = 0$ as the lower limit and the weight of seawater $= W$ at an arbitrary time $= t$ as the upper limit to solve both sides as definite integrals,

$$\int_{1000}^{W} (1)dW = \int_{0}^{t} (-16)\, dt$$

$$W - 1000 = -16t \tag{2.15}$$

we arrive at Eq. (2.12) in just one step.

One kilogram of seawater contains $(1000)(0.035) = 35$ g of salt which will serve as the clue substance because of its unchanged amount over time. The mass fraction of salt, w_S, represents the weight ratio of salt to seawater and is expressed as

$$w_S = \frac{35}{W} = \frac{35}{1000 - 16t} \tag{2.16}$$

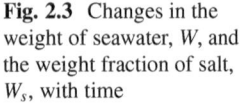

Fig. 2.3 Changes in the weight of seawater, W, and the weight fraction of salt, W_s, with time

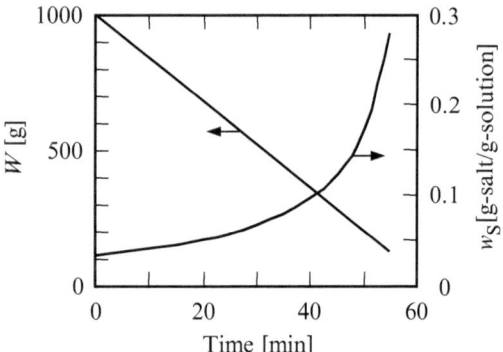

Moreover, Eqs. (2.12) and (2.16) only hold until the saturation solubility of salt (assumed to be sodium chloride) is 28.1 % (w/w) ($t < 54.7$ min). Figure 2.3 shows the changes in the weight of seawater, W, and the weight fraction of salt, w_S, with time. Water evaporates at a constant rate as can be noted from Eq. (2.12); the weight of seawater, W, decreases linearly over time. On the other hand, the weight fraction of salt, w_S, increases only a little at the beginning, but the increment swells markedly as the liquid volume shrinks progressively in the later stage.

The efficiency of the electric heater can be determined as the proportion of energy consumed for evaporation of water out of the total electricity consumption (electrical energy). The output power of the electric heater is given to be 1 kW = 1000 W, and noting that the unit W = J/s, the heater consumes 1000 J/s × 60 s/min = 6.0×10^4 J/min of electrical energy in 1 min. Meanwhile, evaporation of water requires energy, i.e., heat of evaporation. It is given that 16 g of water is evaporated every minute; the heat energy required would be 2.25 kJ/g × 16 g/min = 36 kJ/min = 3.6×10^4 J/min. Thus, the efficiency of the heater at this point is determined to be $(3.6 \times 10^4)/(6.0 \times 10^4) = 0.6$ or 60 %. The result indicates that 40 % of the energy consumed is not utilized for evaporation of water but instead is lost through dissipation to the surroundings of the pot. ◢

Example 2.5 Imagine a container filled with 1 L of a detergent solution with a surfactant concentration of 0.15 g/L. Tap water is being added into the container at 0.4 L/s, while the solution is being simultaneously discharged from the container at the same rate (Fig. 2.4). The solution is assumed to be homogeneously mixed that there are no spatial concentration gradients in the container (known as the perfect mixing condition). How long in seconds will it take for the surfactant concentration to drop below 1.5×10^{-4} g/L which is a thousandth of its initial concentration? And how much tap water will have been added within that period of time?

Solution To solve the problems for the afore-described operation, we need to come up with an equation that describes the change in surfactant concentration, C [g/L] of the detergent solution over time, t [s]. Let the liquid volume in the container be V [L] and the flow rates of both the tap water into the container and the solution out of the

Fig. 2.4 Flow sheet of a dilution operation of a detergent solution

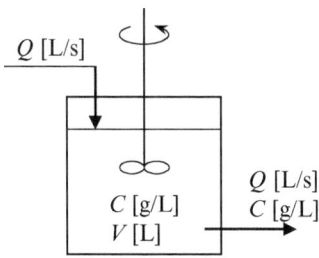

container be Q [L/s]. Considering the mass balance with respect to the surfactant, there is no input of surfactant into but only output from the system, indicating that it is not a steady-state process. Therefore, we need to take into account the term of accumulation in the mass balance equation (left-hand side of Eq. (2.1)). In addition, the surfactant is neither generated nor lost in the container, so the right-hand side of Eq. (2.1) will only consist of input and output.

Let the surfactant concentrations in the container at time t be C and at $t + \Delta t$ (after infinitesimal time has passed since time t) be $C + \Delta C$; the accumulation of surfactant [g] in the container within the time period of Δt is given by $V(C + \Delta C) - VC$. Meanwhile, the input of surfactant into the container during the period of time (Δt) is $Q(0)\,\Delta t = 0$ (the input of tap water does not contain surfactant, so the surfactant concentration is equal to 0), while the output can be described by $QC\Delta t$ (although the surfactant concentration changes from C to $C + \Delta C$ during this period, it can be approximated to C due to the fact that Δt, and therefore ΔC are infinitely small). Substituting these terms in Eq. (2.1) yields

$$V(C + \Delta C) - VC = 0 - QC\Delta t \tag{2.17}$$

Noting that V is constant and dividing the both sides of Eq. (2.17) by Δt gives

$$V\frac{(C + \Delta C) - C}{\Delta t} = -QC \tag{2.18}$$

According to the definition of differential calculus, taking the limit as Δt tends to zero $(\Delta t \to 0)$ gives

$$V\frac{dC}{dt} = -QC \tag{2.19}$$

This is the mass balance equation of the system that we are considering. It is not uncommon that mass balance equations for unsteady-state processes are expressed in the form of differential equations. Equation (2.19) is a variables separable ordinary differential equation. Separating the variables yields

$$\frac{dC}{C} = -\frac{1}{V/Q}dt \tag{2.20}$$

where V/Q is called *average residence time* which has the dimension (unit) of time. Let the surfactant concentration in the container at time $t = 0$ be C_0 (this is the initial condition for Eq. (2.19)); the solution of Eq. (2.19) can be obtained by integrating Eq. (2.20) within the limits:

$$\int_{C_0}^{C} \frac{dC}{C} = -\frac{1}{V/Q} \int_0^t dt$$

$$\ln \frac{C}{C_0} = -\frac{t}{V/Q}$$

$$\frac{C}{C_0} = \exp\left(-\frac{t}{V/Q}\right) \tag{2.21}$$

Taking the logarithm of both sides of Eq. (2.21) and transposing give

$$t = -\frac{V}{Q} \ln\left(\frac{C}{C_0}\right) \tag{2.22}$$

Substituting $C_0 = 0.15$ g/L, $C = 1.5 \times 10^{-4}$ g/L, $V = 1$ L, and $Q = 0.4$ L/s in Eq. (2.22) yields

$$t = -\frac{1}{0.4} \ln\left(\frac{1.5 \times 10^{-4}}{0.15}\right) = 17.3 \tag{2.23}$$

It takes 17.3 s for the surfactant concentration to drop below 1.5×10^{-4} g/L. And $(0.4)(17.3) = 6.92$ L of tap water will have been added through this period of time.

In the example given above, tap water is continuously run into the container until the surfactant concentration drops to 1/1000 of the initial concentration. A common alternative way of reducing the surfactant concentration would be to lade part of the detergent solution out of the container and then top up with the same amount of tap water, and repeat the same operation until the desired concentration is attained. Let us compare the amounts of water needed to achieve the prescribed concentration between the two dilution methods. Let the initial volume of the detergent solution with the initial surfactant concentration of C_0 [g/L] be V [L]. After pouring some of the detergent solution out of the container, the amount of solution left in the container is now v [L]. The remainder of the solution now contains $C_0 v$ [g] of surfactant. When the same amount of tap water is added into the container, the reduced surfactant concentration, C_1, can be described by

$$C_1 = C_0 \left(v/V\right) \tag{2.24}$$

When the same operation is repeated, the further reduced surfactant concentration, C_2, can be described by

$$C_2 = C_1 (v/V) = C_0 (v/V) \cdot v/V = C_0(v/V)^2 \qquad (2.25)$$

And carrying out the same operation for n times would reduce the surfactant concentration to C_n as given by

$$C_n = C_0(v/V)^n \qquad (2.26)$$

Hence, the number of times the operation must be carried out, n, for the surfactant concentration to drop from the initial value of C_0 to C_n can be determined by first transposing Eq. (2.26):

$$(v/V)^n = C_n/C_0 \qquad (2.27)$$

Then taking the logarithm of both sides and rearranging

$$n \log (v/V) = \log (C_n/C_0)$$

$$n = \frac{\log (C_n/C_0)}{\log (v/V)} \qquad (2.28)$$

One may also take the natural logarithm of Eq. (2.27), but in this kind of cases we mostly use the logarithm with base 10 (common logarithm). We will still come to the very same answer regardless of whether the common or natural logarithm is used. ◢

Under identical initial conditions as Example 2.5, if dilution is to be performed by removing 980 mL of the detergent solution from the container and then adding in the same amount of tap water to the remaining 20 mL of detergent solution in the container, how many times has the operation to be performed to attain a surfactant concentration below 1.5×10^{-4} g/L?

Plugging $C_0 = 0.15$ g/L, $C = 1.5 \times 10^{-4}$ g/L, $V = 1$ L, and $v = 0.02$ L into Eq. (2.28):

$$n = \frac{\log \left(1.5 \times 10^{-4}/0.15\right)}{\log (0.02/1)} = 1.77 \qquad (2.29)$$

The dilution operation has to be performed twice for diluting the detergent solution to a thousandth of its initial concentration, and this dilution method uses (0.98) (2) = 1.96 L of tap water which is merely roughly 30 % of the amount needed for the former method. Even if the volume of remainder detergent solution is to be increased by five times to 100 mL in the dilution operation, the operation has only to be repeated twice, and the total amount of tap water needed would be (0.90) (3) = 2.70 L which is still just about 40 % of the amount by the former method.

2.3   Analysis and Presentation of Data

2.3.1   *Logarithmic Manipulation of Data*

$2 \times 2 \times 2$ can be written as 2^3 (2 to the power 3) and 2^3 equals 8. In the other way around, the *logarithm* of 8 to base 2, expressed as $\log_2 8 = 3$, is the number of 2 s we must multiply to get 8 or the *exponent* to which 2 must be raised to produce 8. When $y = a^x$, a must be raised to the exponent of x to produce y and can be expressed as a general equivalent logarithmic statement:

$$\log_a y = x \qquad (2.30)$$

where a is the *base* and y is the *antilogarithm* in base a of x. Moreover, $a^0 = 1$, no matter what value a takes, a must be raised to the exponent of 0 to produce 1. Therefore $\log_a 1 = 0$.

Although the logarithmic base, a, can be any positive number other than 1, 10 and e (=2.718...; the *base of the natural logarithm* or *Napier's constant*) are the bases used in pretty much all cases in the field of food engineering. The use of other bases is rarely encountered. The base-10 logarithm, $\log_{10} y$, is called the *common logarithm* while the base-e logarithm, $\log_e y$, the *natural logarithm* and are often written as

$$\log_{10} y = \log y \qquad (2.31)$$

$$\log_e y = \ln y \qquad (2.32)$$

pH that represents the concentration of hydrogen ion $[H^+]$ is a typical example of a logarithm to the base 10:

$$pH = -\log_{10}\left[H^+\right] = -\log\left[H^+\right] \qquad (2.33)$$

In other words, pH is the negative of the base-10 logarithm of concentration of hydrogen ion (by multiplying the logarithm by -1). $pH = 2$ means the concentration of hydrogen ion, $[H^+]$, is equal to $10^{-2} = 0.01$ mol/L. Meanwhile, the natural logarithm is often found as the solution to the integral of $1/x$ with respect to x:

$$\int \frac{1}{x} dx = \log_e x = \ln x \qquad (2.34)$$

Example 2.6   The difference between $pH = 4.2$ and $pH = 4.6$ is $\Delta pH = 0.4$. What is the difference in terms of concentration of hydrogen ion $[H^+]$?

Solution The concentrations of hydrogen ion at $pH = 4.2$ and $pH = 4.6$ are $[H^+] = 10^{-4.2} = 6.31 \times 10^{-5}$ mol/L and $[H^+] = 10^{-4.6} = 2.51 \times 10^{-5}$ mol/L, respectively, the difference between which is approximately 2.5-folds. ◢

The below equations are some of the logarithmic identities that relate logarithms to one another:

$$\log(xy) = \log x + \log y \quad (\ln(xy) = \ln x + \ln y) \tag{2.35}$$

$$\log(x/y) = \log x - \log y \quad (\ln(x/y) = \ln x - \ln y) \tag{2.36}$$

$$\log x^n = n \log x \quad (\ln x^n = n \ln x) \tag{2.37}$$

Semilogarithmic and double logarithmic graph papers described later are graph papers with logarithmic scales on which exponential data can be graphed directly without having to translate them into logarithms. Here, the logarithm means the common logarithm. Figure 2.5 shows a two-cycle logarithmic scale of a semilogarithmic graph paper. We know that $\log 1 = 0$ and $\log 10 = 1$, let the coordinate of ⓪ in the figure be $x = 1$, and then the coordinate of ① becomes $x = 10$. Once the coordinates of $x = 1$ and $x = 10$ are set, from the logarithmic identity expressed by Eq. (2.36), we also know that $\log 100 = \log 10^2 = 2\log 10 = 2$, so ② which is located twice as far from ⓪ as ① will be $x = 100$. Moreover, on commercially available semilogarithmic graph papers, you can see that the line at ⓐ located 1.89 cm from ⓪ is slightly bolder than those in between it and ⓪. The distance between ⓪ and ① is 6.29 cm while that between ⓪ and ⓐ is 1.89 cm, making the length ratio $1.89/6.29 = 0.300$ which is nearly equal to $\log 2 = 0.301$. That is to say that letting the distance between ⓪ and ① be 1 unit length, the tick mark of ⓐ will indicate $\log 2$. Likewise, ⓑ is 3.00 cm away from ⓪ and thus indicates $3.00/6.29 = 0.477 \approx \log 3$, representing $x = 3$. Meanwhile ⓒ is also 1.89 cm to the right of ① and indicates $\log 2$ as well. Hence, the distance between ⓪ and ⓒ consists of 1 unit length from ⓪ to ① and $\log 2$ from ① to ⓒ. From the logarithmic identities expressed by Eqs. (2.37) and (2.35), $1 + \log 2 = \log 10 + \log 2 = \log(10 \times 2) = \log 20$. So ⓒ represents $x = 20$

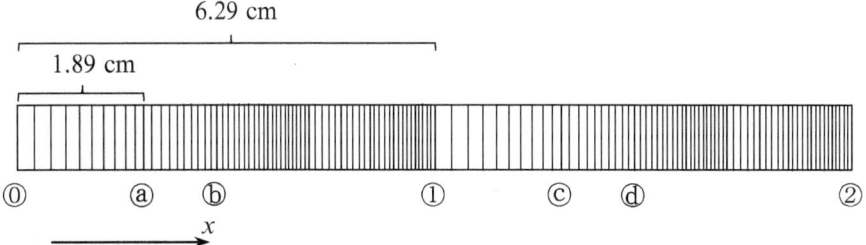

Fig. 2.5 Logarithmic scale

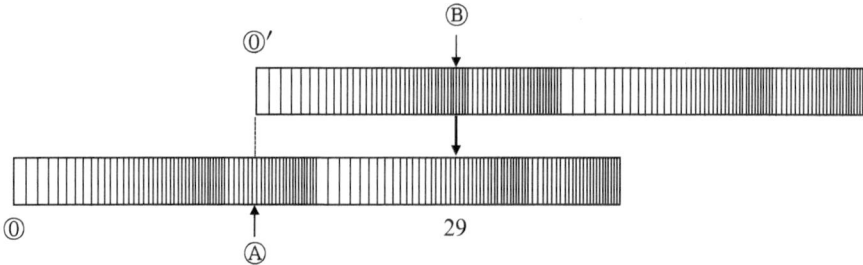

Fig. 2.6 Multiplication operation based on logarithmic identities

(instead of 11) and ④ represents likewise $x = 30$. The above-described spacings that indicate the logarithmic distances are known as the *logarithmic scale*.

By leveraging the logarithmic identities discussed above, multiplication and division can be performed easily using logarithmic graph papers. We will demonstrate the calculation of 63×46 as an example. As 63×46 can be transformed to $63 \times 46 = 6.3 \times 10 \times 4.6 \times 10 = 6.3 \times 4.6 \times 10^2$, we will just have to calculate 6.3×4.6. Ⓐ in the lower logarithmic scale in Fig. 2.6 represents 6.3, indicating the length of log6.3 from ⓪. Meanwhile, Ⓑ in the upper logarithmic scale represents 4.6, indicating the length of log4.6 from ⓪′. Aligning ⓪′ of the upper scale with Ⓐ in the lower scale and reading the value on the lower scale at the location of Ⓑ give the value of approximately 29. That is $63 \times 46 = 6.3 \times 4.6 \times 10^2 \approx 29 \times 10^2 = 2.9 \times 10^3$. This multiplication operation is performed on the basis of the logarithmic identity (Eq. (2.35)) that states that $\log 6.3 + \log 4.6 = \log(6.3 \times 4.6)$. Perhaps you may think that the answer is inaccurate because $63 \times 46 = 2898$. Even one tries to read values by eye between the tick marks for the minor units on the logarithmic scale of a graph paper, one can only retain 2 or the most 3 significant figures. Therefore, the values resulted from the addition of those numbers will also have the same number of significant figures which is 2 in this example. You may feel that the method is devoid of accuracy but 2 or 3 significant figures are usually adequate in engineering calculation. Organizing empirical data using spreadsheet software and then processing with operation such as division may result in data with 8 or more significant figures which are completely meaningless because the actual measured data obtained through experiments mostly are not recorded up to that number of significant figures. In fact, reporting experimental results with inappropriately high numbers of significant figures not only implies overestimated precisions of the experiments but also gives away one's lack of understanding in the use of significant figures.

Note that we can also perform division, the inverse of the above-described multiplication, based on Eq. (2.36). In addition, Eq. (2.37) states that $\log \sqrt{x} = \log x^{1/2} = (1/2) \log x$, so the logarithmic scale can also be manipulated to find the square root of a number. For instance, $\sqrt{67}$ can be determined by first plotting 67 on the logarithmic scale, then folding that tick mark of 67 into that of 1, and finally reading the scale along the fold line to be approximately 8.2. These manipulations are the principle of calculation for slide rule, a calculation tool commonly used prior to the advances of the electronic scientific calculator and computer.

2.3.2   How to Draw a Graph

In many cases, actual measured data collected from experiments and observations and their analysis results are obviously easier to understand when they are translated into self-explanatory visual forms. Although data are converted into graphs using computer software in almost every situation nowadays, being able to draw a decent graph by hand is still fundamental. Further, it also helps one comprehend the various options available in the computer software. To draw a graph by hand, we first need a piece of graph paper. Apart from the *normal graph paper* that uses a linear scale for both axes on which every division has the same spacing, there are also other special graph papers (semilogarithmic graph paper, double logarithmic graph paper, and normal probability paper) available in the market. In this section, we will walk you through on how to use the normal, *semilogarithmic*, and *double logarithmic graph papers* as they are likely to be of more frequent use in the future. The caption of a graph shall always be placed below the graph, whereas the title of a table shall always be on top. Although conditions and annotations are often stipulated below the caption, they can sometimes be found in the graph as well.

(a) *Normal Graph Paper*: The normal graph paper with a linearly graduated scale on both axes is probably the most frequently used one of all the different types of graph papers. When we draw a graph, there are a few aspects we need to include (e.g., the axes, the axis titles, and the legend of symbols), and these aspects can be presented in several different ways.

Example 2.7 The cooking time and the corresponding moisture content were recorded through a cooking operation of spaghetti with different thicknesses in boiling water as tabulated in Table 2.5. Plot the measured data on a normal graph paper.

Solution A graph may consist of one abscissa (horizontal axis) and an ordinate (vertical axis) as depicted in Fig. 2.7a, or it may also be drawn with two horizontal axes, one at the bottom and the other on top, and two vertical axes, one on each side, such that these pairs of axes enclose the plot area of the graph as illustrated in Fig. 2.7b. Reference for points on the scales on the axes can be provided either

Table 2.5 Water absorption of spaghetti with different thicknesses

Cooking time [min]		0	5	10	15	20	25	30
Dry basis moisture content [g-water/ g-d.m.]	Thin	0.19	1.38	2.08	2.93	3.29	3.79	4.18
	Thick	0.19	1.01	1.57	2.08	2.52	2.93	3.26

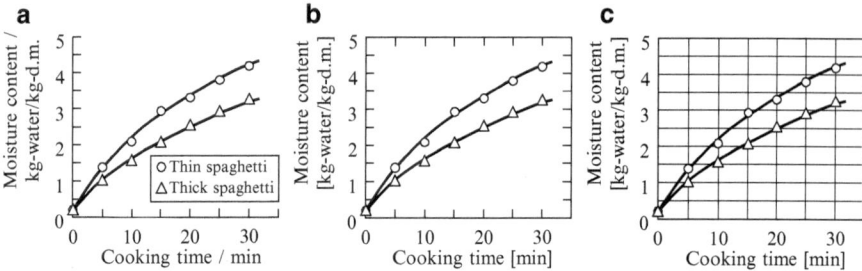

Fig. 2.7 Various ways for constructing a graph taking a cooking operation of a thin spaghetti (○) and a thick spaghetti (△) for example

by the use of tick marks as shown in Fig. 2.7a, b or gridlines that stretch all the way across the bordered plot area as depicted in Fig. 2.7c. Subsequently, we label the tick marks and title the axes so that the readers know what the axes represent. The physical quantities presented on the axes usually have units. The units of the physical quantities presented can either be stipulated following a slash (/) (Fig. 2.7a) or parenthesized at the end of the axis titles (Fig. 2.7b, c). In the cases where the physical quantities are dimensionless (quantities without units), the axis titles may be either left just as they stand or ended with a parenthesized hyphen ([-]). The title of the vertical axis shall be written bottom to top such that it can be read correctly from the right. When different groups of data are presented in one graph, each group of data shall be represented by a symbol of specific shape (○, △) or color that distinguishes the group from the others. A special text object called the graph legend which provides explanatory information of the symbols may be created in the plot area (often on the right) of the graph to allow readers to identify the data from each group (Fig. 2.7a). Otherwise, the descriptions of the symbols may be included in the figure caption as in that of Fig. 2.7. Meanwhile, the symbols of different shapes (○, △) shall be drawn accurately using a *stencil template* (Fig. 2.8) instead of freehand. Once the data have been plotted, we draw a smooth trend line connecting all the data points or a theoretically calculated curve using a French curve (Fig. 2.9) or a flexible curve instead of freehand. ◢

(b) *Semilogarithmic Graph Paper*: Suppose we have a data set of which the variable y obeys the below exponential relationship:

$$y = ae^{bx} \tag{2.38}$$

where a and b are constants. So we need to determine a and b that fit the data set to calculate the approximate curve. Taking the logarithm of both sides gives

$$\log y = \log a + (b \log e) x \tag{2.39}$$

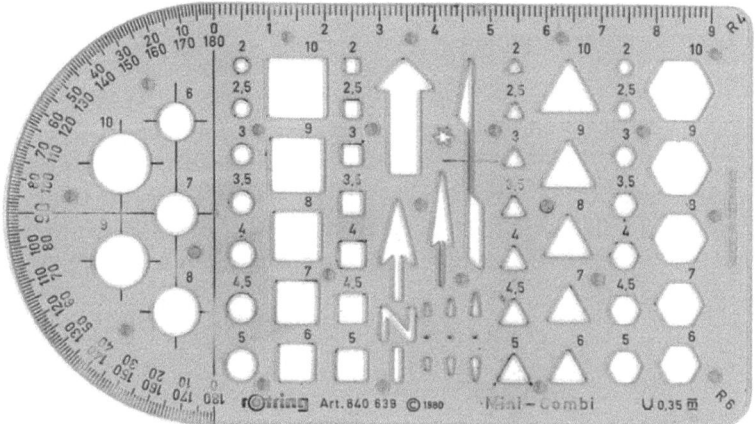

Fig. 2.8 A stencil template

Fig. 2.9 French curves

Hence, plotting log y as a function of x will yield a straight line of y-intercept log a and slope (*b* log *e*).

Although we may translate the data into logarithms using a calculator and then plot the translated data on a normal graph paper, there are several types of special graph papers available in the market, which may suit our needs and thus save us the trouble of converting the data. In this case, the semilogarithmic graph paper that uses a linear scale for one axis and a (common) logarithmic scale for the other axis will serve the purpose. Note that the slope of the straight line produced on this type of graph paper cannot be obtained from the ratio of the lengths of legs of the triangle measured using a ruler (refer to Sect. 6.3 for details).

(c) *Double Logarithmic Graph Paper*: Now, let us say we have a data set of which the relationship between variables x and y can be described by

$$y = ax^b \tag{2.40}$$

Again, taking the logarithm of both sides gives

$$\log y = \log a + b \log x \qquad (2.41)$$

where a and b are constants. Hence, plotting log y versus log x will give a straight line of slope b. Meanwhile a can be determined by reading the coordinate of an arbitrary point on the straight line and then plugging the values into Eq. (2.40). In this case, plotting y against x directly on a double logarithmic graph paper would save us a lot of work without us having to find the logarithms of y and x with a calculator.

Exercise

2.1 Convert the non-SI units of the following quantities to SI units.
   (a) Gas constant, R, 0.0821 L · atm/(mol · K); (b) latent heat of melting of ice, 79.7 cal/g; (c) vapor pressure of water (25 °C), 0.0323 kgf/cm^2; (d) viscosity of water (25 °C), 8.90×10^{-3} g/(cm · s); and (e) thermal conductivity of water (25 °C), 0.00145 cal/(cm · s · °C).

2.2 What are the wet basis and dry basis moisture contents of the fresh spaghetti made by the kneading process described in Example 2.2?

2.3 Dried spaghetti weighing 19.2 kg with a dry basis moisture content of 11 % was made from 20 kg of durum semolina that contains 14 % moisture on wet basis. How much out of the raw material is recovered as finished product (what is the yield)?

2.4 Ten kilograms of strawberries with a wet basis moisture content of 91 % is added with 5.0 kg of granulated sugar and cooked down to prepare a strawberry jam that contains 45 % moisture on wet basis. What is the weight of the finished product of strawberry jam? Granulated sugar contains negligible amount of moisture, (0.02 %) and thus assume the moisture content to be 0.

2.5 A drying operation is known to be able to remove 80 % of the initial moisture in a food product that contains 70 % (w/w) moisture. How much water (mass) will be evaporated if 1 kg of this food product in its initial condition is to be subjected to the same drying operation? Determine the composition (weight ratios of solid and water) of the dried food product.

2.6 Ten thousand kilograms (1.0×10^4 kg) of sugarcane solution with a sugar concentration of 38 % (w/w) are concentrated in an evaporator to a concentration of 74 % (w/w) on a daily basis. How much sugarcane solution is produced, and how much water is evaporated in the unit of kilogram every day?

2.7 Tapioca starch, a raw material of noodles, confectionery, and other similar food products, is produced by drying the starch granules of cassava root from the wet basis moisture content of 66 % (w/w) to 5 % (w/w) and then pulverizing the dried granules. How much starch granules in the units of kg/h needs to be dried to produce 5000 kg/h of tapioca starch? Also, how much water is evaporated by this drying process?

2.8 Ten thousand kilograms $(1.0 \times 10^4$ kg) of saturated solution of Na_2CO_3 is prepared at 30 °C. To what temperature has one to cool down the solution from 30 °C for 3.0×10^3 kg of $Na_2CO_3 \cdot 10H_2O$ crystals (the crystals do not contain water in liquid form) to crystallize out? The aqueous solubility of Na_2CO_3 is shown in Table 2.6.

2.9 A hundred tons of cellulose and 1000 kg of microorganisms are supplied to a wastewater treatment equipment on a daily basis, while 10 t of cellulose and 15,000 kg of microorganisms are discharged daily. The digestion speed of cellulose by the microorganisms is 10^4 kg/day. The microorganisms proliferate at the speed of 2×10^4 kg/day, while they perish due to lysis at the rate of 5×10^2 kg/day. Estimate the accumulation of cellulose and microorganisms within the wastewater treatment equipment.

2.10 Ten thousand kilograms $(1.0 \times 10^4$ kg) of soybeans are processed into soybean oil and soybean flakes through a three-step treatment as illustrated in Fig. 2.10. The soybeans have a composition of 34 % (w/w) protein, 27 % (w/w) carbohydrates, 10 % (w/w) fibers, 11 % (w/w) water, and 18 % (w/w) oil. In the first step of treatment, the soybeans are pulverized and pressed to separate the oil resulting in pressed oil and soybean flakes that contain 6.5 % (w/w) of oil. Assume that the pressing operation manages to separate only the oil component from the soybeans but not the other components. In the next step of treatment, the soybean flakes are subjected to solvent extraction using hexane, reducing the oil content of the soybean flakes to 0.4 % (w/w) while yielding oil-containing hexane. Assume here that there is no residue of hexane in the post-extraction soybean flakes. Finally, the flakes are dried to the moisture content of 7 % (w/w). Determine the following: (a) the weight of the soybean flakes after the first step of treatment, (b) the weight of the solvent-extracted soybean flakes after the second step of treatment, and (c) the weight of the dried soybean flakes after the third step and the protein composition.

Table 2.6 Solubility of Na_2CO_3

Temperature [°C]	0	10	20	30
Solubility [g–Na_2CO_3/100 g–H_2O]	7.0	12.5	21.5	38.8

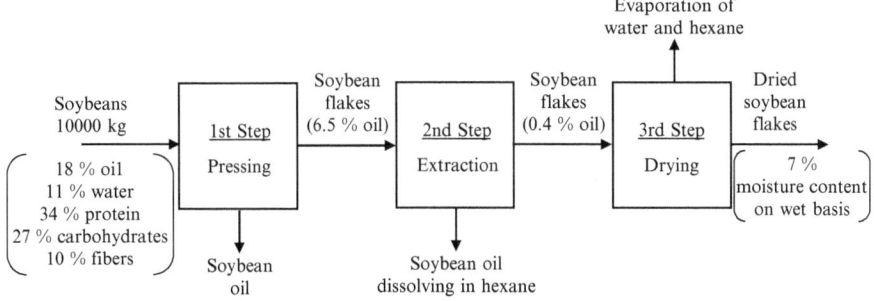

Fig. 2.10 Three-step treatment of soybean oil processing

2.11 A thousand grams of water is added to 450 g of starch that initially contains 13 % water on wet basis. α-Amylase and glucoamylase (both enzymes) are added to the mixture to catalyze the hydrolysis reaction of starch, and 98 % (with reference to the glucose residues that make up the starch molecules) of the starch are broken down to glucose. What are the ratios (by weight) of the unhydrolyzed starch, water, and glucose in the sugar syrup obtained. Let the molar mass of the glucose residues that make up the starch molecules be 162 g/mol and the terminal glucose residues be also of the same molar mass.

2.12 Alcohol concentration is often presented in volume percent [% (v/v)]. What is the ethanol concentration in weight percent [% (w/w)] (or [wt%]) of a 12 % (v/v) ethanol solution obtained from fermentation? Let the respective densities of ethanol and water be 0.79 and 1.00 kg/L.

The ethanol solution is separated continuously using a fractionating column (a piece equipment that leverages the differences in volatilities of the constituents of a liquid mixture for separating it into its component parts) into two ethanol solutions with respective ethanol concentrations of 20.8 and 2.5 % (w/w). The low-boiling-point fraction with the higher (ethanol) concentration is called the distillate, while the high-boiling-point fraction with the lower concentration is called the bottoms. If the 12 % (v/v) ethanol solution is supplied to the fractionating column at the rate of 100 kg/h, what are the respective output rates [kg/h] of the distillate and bottoms?

2.13 A mayonnaise is prepared by drizzling vegetable oil (cooking oil) into an aqueous phase consisting mainly of egg yolks and vinegar while vigorously agitating the mixture. The ratio (by weight) of vegetable oil to egg yolk to vinegar (including salt, etc., hereinafter called vinegar) is roughly 70:15:15. If we are to prepare a mayonnaise at this weight ratio with ten eggs, how much vegetable oil and vinegar do we need? Let each egg weigh 60 g and the egg yolk 20 g. At the end of the preparation process, 1250 g of mayonnaise is collected, while the rest sticks in the mixing bowl. What is the weight ratio of the unrecoverable mayonnaise?

2.14 Determine the pH values of hydrochloric acid solutions with concentrations of 0.01 and 0.2 mol/L. Let the degree of dissociation of hydrogen chloride be 1.

2.15 What is the pH of an equivolume mixture of hydrochloric acid solutions of pH 2.8 and pH 4.2? Let the degree of dissociation of hydrogen chloride be 1.

2.16 A sucrose solution of volume V [cm^3] consists of m_W [g] of water and m_S [g] of sucrose. Assuming that the volume of the solution can be expressed as the sum of the respective volumes of sucrose and water, the relationship described by Eq. (2.42) thus holds

$$V = \frac{m_S}{\rho_S} + \frac{m_W}{\rho_W} \qquad (2.42)$$

where ρ_S [g/cm^3] and ρ_W [g/cm^3] are the densities of sucrose and water, respectively. Table 2.7 shows the volumes of the sucrose solutions prepared by solubilizing various quantities of sucrose in 100 g of water at a constant temperature. Find out the densities of sucrose and water.

Table 2.7 Weight of sucrose, m_S, and the volume of its solution, V

m_S [g]	35	70	110	150	180
V [cm³]	122.3	144.4	169.5	194.6	213.5

Table 2.8 Variation of absorbance at 450 nm of a dye solution

t [min]	0	5	10	15	20	25
A		0.854	0.469	0.257	0.141	0.077	0.043

Table 2.9 Viscosity of ethanol at different temperatures

Temperature [°C]	−20	0	20	40	60
Viscosity [mPa·s]	2.82	1.77	1.20	0.834	0.592

Table 2.10 Adsorption of color pigments by activated charcoal

C [mg/L]	2.1	3.9	9.8	19.8	28.8
q [g/g]	2.56	3.18	4.19	5.31	5.91

2.17 A 1.0-m³ tank contains water stained with a yellow dye. Clean water is supplied into the tank at the flow rate of Q [m³/min], while the content is being discharged at the same rate. The content of the tank is measured over time for absorbance, A, at 450 nm, and the actual measured data are presented in Table 2.8. When the absorbance, A, is proportionate to the dye concentration, the relationship between time, t, and absorbance, A, can be expressed in a similar manner as Eq. (2.21) (refer to Example 2.5) by

$$\frac{A}{A_0} = \exp\left(-\frac{t}{V/Q}\right) \tag{2.43}$$

where A_0 is the initial absorbance and V is the volume of liquid in the tank. Determine the flow rate, Q, using a semilogarithmic graph paper.

2.18 The viscosity, μ [Pa·s], of a fluid at temperature, T [K], can be approximately expressed by

$$\mu = be^{a/T}$$

Table 2.9 summarizes the viscosity data of ethanol at various temperatures. Determine the parameters a and b by the use of a semilogarithmic graph paper.

2.19 Activated charcoal is added to a colored sugar syrup to adsorb the color pigments. The color pigment concentration, C [g-color pigment/L-solution], and the amount of color pigment adsorbed, q [g-color pigment/g-activated charcoal], at equilibrium are shown in Table 2.10. When the relationship between C and q is described by the Freundlich equation as expressed by Eq. (2.44), determine the parameters a and b using a double logarithmic graph paper.

$$q = aC^b \tag{2.44}$$

Chapter 3
Wheat Flours and Their Derived Products

Abstract We will discuss the following items by taking wheat flour and food made from wheat flour as examples. First, we will elaborate on the size and particle-size distribution of particles of irregular shapes like wheat flour particles. Then, we will provide an account for stress-strain curves that express the relationships between the force (stress) applied onto food materials and their deformation (strain). Further, we will explain the definitions of and the difference between moisture content and water activity in food materials. Besides, we will cover moisture sorption isotherms which express the relationship between humidity (water activity) and moisture content of food products and also go into detail on hot air-drying. The properties and storage stability of food products may differ as they may either be in the glassy or rubbery state depending on the moisture content and temperature. Finally, we will touch on the glass transition of food products which refers to the transition between the glassy and rubbery states.

Keywords Characteristic diameter • Particle-size distribution • Texture • Stress-strain curve • Moisture content • Water activity • Water sorption isotherm • Hot air-drying • Glassy state • Rubbery state • Glass transition

3.1 Classification of Wheat Flours and Their Applications

We have enumerated several foods in Chap. 1 that are made of or contain wheat flour like bread, spaghetti, tempura, and so on. Wheat grains, as the raw material of wheat flour, are classified not only by cultivars but also by various aspects such as cultivation periods, and properties. The wheats that are sown in autumn and harvested in summer are known as winter wheats, whereas those sown in spring and harvested in autumn are called spring wheats. Further, wheat grains are also categorized by hardness of grain into hard and soft wheats. Besides, there are also red wheats which contain red or reddish-brown pigments as opposed to white wheats which are lower in pigment content. Durum wheats used in making pasta, though lightly pigmented, belong to a class of hard white wheats that are harder than the normal hard wheats. Owing to the hardness, durum wheats are not milled into flour but instead ground into course middlings regarded as semolina for use.

© Springer Science+Business Media Singapore 2016
T.L. Neoh et al., *Introduction to Food Manufacturing Engineering*,
DOI 10.1007/978-981-10-0442-1_3

Table 3.1 Types of wheat flours and their main applications

Type	Protein content [%]	Main applications
Strong flour (bread flour)	10.7–13.0	Bread loaves
Semi-strong flour (all-purpose flour)	10.5–12.0	Bread rolls, Chinese noodles
Medium flour (all-purpose flour)	8.0–10.0	French loaves, snacks, and noodles
Weak flour (cake flour)	6.5–8.0	Snacks, tempura, and cakes

There are extremely diversified types of wheat flours being tailored to the wide-ranging uses in an abundant variety of foods. Generally, wheat flours are classified into strong (bread flour), semi-strong (all-purpose flour), medium (all-purpose flour), and weak (cake flour) flours. In Japan, wheat flours are also graded according to mineral (ash) content into first-grade (roughly 0.3–0.45 %), second-grade (0.45–0.65 %), third-grade (0.65–1.0 %), and low-grade (1.0–2.0 %) flours. Basically wheat flours of first or second grade are used in food processing. Table 3.1 summarizes the applications of wheat flours with different type-grade combinations in various kinds of foods. The protein content differs widely from type to type as well, depending on reference source. Even for bread making alone, different types of flours are utilized for different varieties of breads, for example, strong flours are used for sandwich loaves, semi-strong flours for bread rolls, and medium-strength flours for French loaves. Japanese wheat noodles or udon are made from medium-strength flours, while "tempura" and "okonomiyaki" (a thin pancake cooked on a hot plate with chopped cabbages and bits of meat or seafood) are made from weak flours.

3.2　Particle Size of Wheat Flours

A great number of food ingredients, just like wheat flour and sugar, are in powder form. Even by the naked eye, we could observe that the small particles are various in shape instead of being all spherical. The particle-size measurement and the average particle size of those particles irregular in shape would be further discussed in this section together with the equations that describe the particle-size distribution.

3.2.1　Characteristic Diameter

Wheat flours are obtained by first *crushing* wheat grains and then *pulverizing* them into powder form. The combined process is regarded as *milling*. The wheat flour will subsequently be sifted to uniformize the particle size to some extent (Fig. 3.1). Wheat flour owes its irregularity in particle shape to the above-described process operations by which it is produced (Fig. 3.2). There are a number of ways to represent the size (diameter) of particles of irregular shapes. In this subsection, we

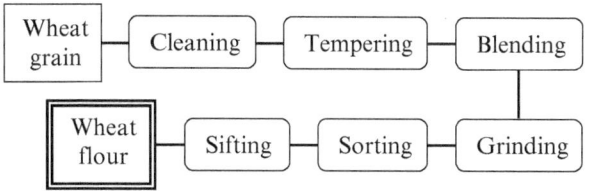

Fig. 3.1 Process outline for manufacturing of wheat flour

Fig. 3.2 Scanning electron micrograph of a wheat flour sample

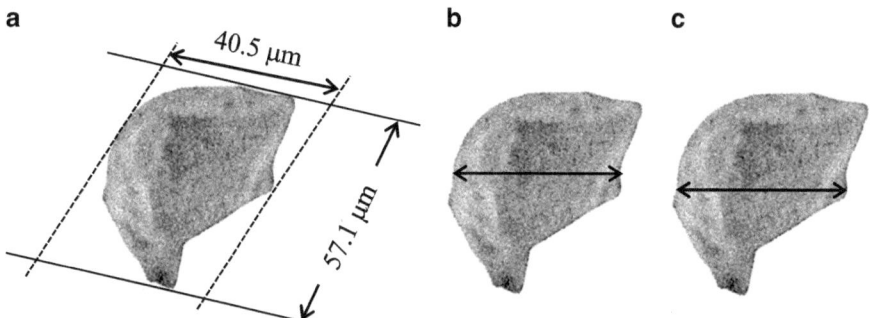

Fig. 3.3 (**a**) Feret diameter, (**b**) Martin diameter, and (**c**) Krumbein diameter of a wheat flour particle

will give a detailed account of the methods for the determination of particle size using two-dimensional images such as micrographs. Another method of determining particle size using sieves, which is widely employed in the food industry, will be detailed in the subsequent Sect. 3.2.2.

The distance between two parallel planes restricting a particle along a specified direction perpendicular to that direction is defined as the *Feret* (or Green) *diameter*. In Fig. 3.3a, the particle diameter determined by the parallel solid lines is 57.1 μm, whereas the particle diameter reads 40.5 μm between the dashed lines. The readings of diameter differ substantially depending on the direction in which the lines are drawn and also the orientation of the particles. Nonetheless, the influence of these two factors on average particle diameter becomes negligible because at least hundreds of particles are to be measured by this method. There are also several other alternative measures of diameter of irregularly shaped particles based on the projected areas of particles. The length of line that bisects the projected area of an irregular particle in a specified direction is known as the *Martin diameter* (Fig. 3.3b). The *Krumbein diameter* refers to the longest line of a specific direction across the projected area of an irregular particle (Fig. 3.3c). The diameter of the circle with the same area as the projected area of an irregular particle is named *projected area equivalent diameter* (or the Heywood diameter). The diameter of the circle with a circumference of the same length as the perimeter of the projected area of an irregular particle is called *equi-circumferential circle equivalent diameter*. Besides, there are still other methods for estimating a characteristic diameter. For instance, the Stokes diameter of a particle is calculated from its terminal settling velocity based on the fact that particles of certain sizes are known to fall or settle in gaseous and liquid (collectively known as fluids) media at certain constant velocities (terminal velocities).

Gravitational force acts in vertically downward direction on a spherical particle measuring d in diameter that settles through a fluid medium while buoyancy and drag force (resistance) by the fluid act in the exact opposite direction (Fig. 3.4). Within a short period of time after starting to free-fall from stationary condition, the particle achieves a constant velocity (zero acceleration) when the sum of the buoyancy and the drag force equals the gravitational force. This highest constant velocity attainable by a particular object falling through a particular fluid medium is known as the *terminal velocity*, u_t which, in the case of fine particles, is given by

$$u_t = \frac{(\rho_p - \rho_f)\, gd^2}{18\mu} \quad \text{or} \quad d = \sqrt{\frac{18u_t\mu}{(\rho_p - \rho_f)\, g}} \tag{3.1}$$

where ρ_p [kg/m^3] is the particle density, ρ_f [kg/m^3] is the density of the fluid, μ [Pa·s] is the fluid viscosity, and g [m/s^2] is the gravitational acceleration. The diameter of a particle computed by this equation from its terminal velocity is known

Fig. 3.4 Forces acting on a particle in a fluid

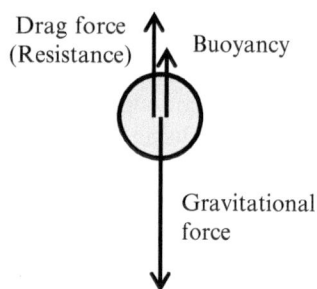

as the *Stokes diameter*. In addition, the fluid viscosity may also be estimated from the terminal velocity of a spherical particle with a known diameter.

Example 3.1 What is the terminal velocity of a spherical water droplet of 0.04-mm diameter free-falling through air from 1000-m altitude? Assume the density, ρ_f, and viscosity, μ, of air to be 0.0012 g/cm³ and 1.8×10^{-4} g/(cm·s), respectively. Compare the terminal velocity with the falling velocity of an identical water droplet under vacuum condition (zero air resistance).

Solution Standardizing the quantities according to the International System of Units (SI) gives $d = 4 \times 10^{-5}$ m, $\rho_f = 1.2$ kg/m³, and $\mu = 1.8 \times 10^{-5}$ kg/(m·s). And we assume the density of the water droplet, ρ_p, to be 1000 kg/m³. By substituting these values into Eq. (3.1), the terminal velocity could be calculated as follows:

$$u_t = \frac{(\rho_p - \rho_f)\, g d^2}{18\mu} = \frac{(1000 - 1.2)\,(9.8)\left(4 \times 10^{-5}\right)^2}{(18)\left(1.8 \times 10^{-5}\right)} = 4.8 \times 10^{-2} \text{ m/s}$$

Technically, we would need to validate the application of Eq. (3.1) by this calculated value of terminal velocity.

Next, the velocity of a water droplet free-falling across the distance of x under vacuum condition may be represented by

$$v = \sqrt{2gx} \tag{3.2}$$

If we plug in 9.8 m/s² and 1000 m for g and x, respectively, the velocity of the identical water droplet would be

$$v = \sqrt{(2)(9.8)(1000)} = 140 \text{ m/s}$$

which is equivalent to 504 km/h (approximately 1.7 times the maximum speed of the "Nozomi Shinkansen" bullet train (300 km/h)), whereas the resistance caused by air is remarkably intense that the terminal velocity would be a mere 0.17 km/h (4.8×10^{-2} m/s). This explains why raindrops are not falling onto the Earth at an extremely high speed that might even break through an umbrella. ◢

Example 3.2 What are the Martin diameter and the Krumbein diameter of the wheat flour particle illustrated in Fig. 3.3?

Solution Since Martin diameter refers to the length of the bisector of the projected area of an irregular particle in a specified direction, the projected area of the particle is supposed to be calculated for accurate determination of its Martin diameter. Nonetheless, by drawing the projected area bisector viscerally, it is determined to be 41.8 μm in length (Fig. 3.3b). Furthermore, by measuring intuitively the longest distance across the projected area, the Krumbein diameter is determined to be 42.3 μm (Fig. 3.3c). ◢

3.2.2   Particle-Size Distribution

It is hard to give someone a good grasp of the particle-size distribution of a particular granular (powder) sample by just listing the vast number of diameter readings measured of that sample. This piece of information would definitely be more easily understandable if presented graphically. There are two ways to present particle-size distribution by the use of bar graphs (histograms) and line graphs. One of the ways is to present the particle size on the horizontal axis and the ratio (or frequency) of number of particles in certain size classes on the vertical axis. In this method, the particles are normally grouped discretely into defined size ranges. The distribution presented in this manner is referred to as a *frequency distribution* or a *density distribution* based on the number of particle sizes. The other way would be to present the particle size on the horizontal axis, but instead of the ratio of particle-size range groups, the ratio of number of particles greater (or smaller) than certain sizes is presented on the vertical axis. This type of distribution is known as a *cumulative distribution* based on the number of particle sizes.

Example 3.3 The particle size (Feret diameter) of a wheat flour sample measured under microscope is grouped under appropriate size ranges and tabulated in Table 3.2. Particles of less than 10-μm diameter were not observable. Particles greater than 210 μm were not present as well because they were removed during the manufacturing process. Construct graphs that show the number-based frequency distribution and also the number-based cumulative distribution which represents the ratio of particles with a diameter greater than a specific size.

Solution Plotting on a normal graph paper the graph with the horizontal axis corresponding to particle size and the vertical axis corresponding to the ratio of number of particles within the size ranges given in Table 3.2 to the total number of particles (frequency), we would obtain a graph similar to that illustrated in Fig. 3.5a. Generally, common logarithmic scales are used for the horizontal axis as the particles often measure from several to a few hundred microns in diameter. Figure 3.5b shows the frequency distribution plotted with a logarithmic scale on the horizontal axis. The type of graph papers with an arithmetic (linear) scale for one axis and a logarithmic scale for the other is called semilogarithmic graph papers.

Table 3.2 Particle-size distribution of a wheat flour sample measured under microscope

Particle size [μm]	Quantity	Fraction	Cumulative value (integration value)
10–20	331	0.623	0.999
20–40	134	0.252	0.376
40–60	41	0.077	0.124
60–80	15	0.028	0.047
80–100	7	0.013	0.019
100–210	3	0.006	0.006
Total	531		

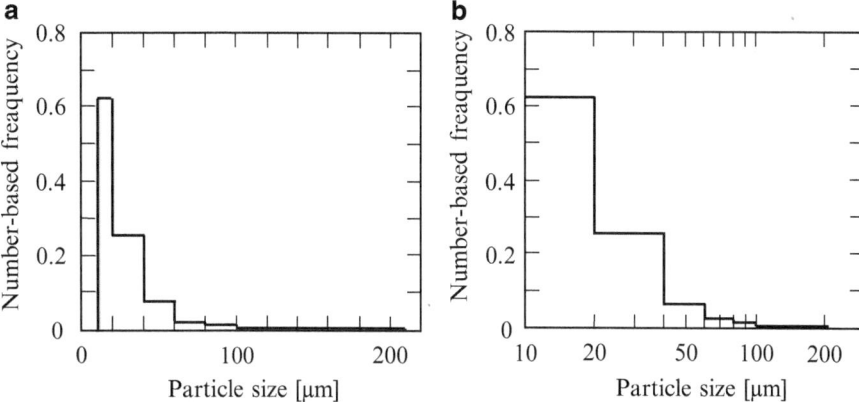

Fig. 3.5 Number-based frequency distribution of particle size of wheat flour. Horizontal axis: (**a**) normal scale and (**b**) logarithmic scale

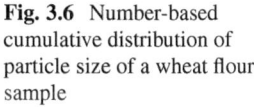

Fig. 3.6 Number-based cumulative distribution of particle size of a wheat flour sample

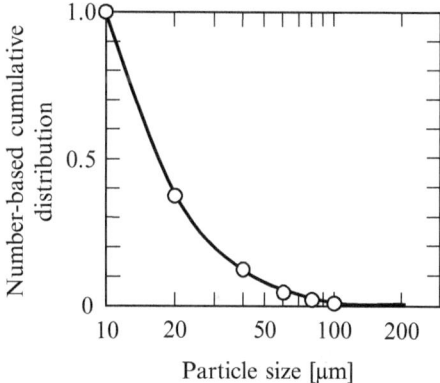

Next, if you sum up the frequencies of particles in the particle-size range groups one by one in descending particle-size order, you will obtain the ratio of particles with diameters greater than the lower limit of each size range as summarized in the fourth column of Table 3.2. The sum of the frequencies of all size groups is 0.999 as tabulated in Table 3.2 instead of 1.000 which it is supposed to be because the frequency data have been rounded to three digits after the decimal point during calculation. Plotting the integration values against particle size on a semilogarithmic graph paper gives Fig. 3.6. This is the so-called number-based cumulative distribution. ◢

Other than the undoubtedly troublesome method of measuring particle size under microscope, there are a few other means for the determination of particle-size distribution based on different measurement principles. The use of *sieves* (Fig. 3.7) with square meshes made of stainless steel or synthetic fibers is the most widely employed means. The distance between two adjacent strands on a mesh is known

Fig. 3.7 (**a**) Sieve and (**b**)
opening of sieve

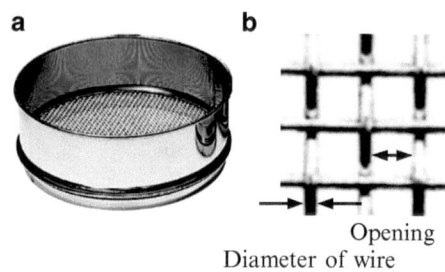

Opening
Diameter of wire

as the aperture (opening) through which particles of sizes smaller than that would
fall. The larger particles would be retained on the mesh. By sifting a powder sample
through a series of sieves with different aperture sizes, the sample can be separated
by range of particle size. The described operation is regarded as *classification*. The
aperture size is often presented in the unit of "*mesh*" which is defined as the number
of openings per inch (unit symbol: in) of mesh. Therefore, the greater the mesh
count is, the smaller the aperture would be. Inch is one of the units of length
in the imperial system (also known as British imperial). An inch is equivalent
to 25.4 mm. The length of a mesh includes the diameters of the strands and the
apertures. The stainless steel strands of a standard 100-mesh sieve are standardized
such that they measure 0.104 mm in diameter, thus giving the sieve an aperture size
of $(25.4/100) - 0.104 = 0.150$ mm $= 150$ μm.

By classifying a powder sample through a series of sieves with different aperture
sizes and determining the weight of powder retained on each sieve, the frequency
distribution of the sample with reference to weight can be obtained. Dividing the
weight of a sample that passes through a sieve by the total sample weight gives the
ratio (weight fraction) of particles with a diameter smaller than the aperture size,
which is known as *cumulative passage fraction* (undersize integration). Meanwhile,
the weight fraction of the particles retained on the mesh represents the ratio of particles of larger sizes than the aperture size, which is known as *cumulative retention
fraction* (oversize integration). Both fractions represent cumulative distributions and
their relation could be expressed as

$$\text{Cumulative passage fraction} = 1 - \text{cumulative retention fraction} \qquad (3.3)$$

Curves obtained in the plots of cumulative passage fraction and cumulative
retention fraction (on an arithmetic scale) versus particle size (aperture size of sieve)
(normally on a logarithmic scale) are regarded, respectively, as a passage fraction
(undersize) curve and a retention fraction (oversize) curve.

Example 3.4 A wheat flour sample was classified using seven sieves of different
aperture sizes stacked on top of the other. The weights of wheat flour retained on
each sieve and that passed the 26-μm sieve are recorded as shown in Table 3.3.
Create graphical representations of (1) the frequency distribution based on weight,
(2) the passage fraction curve, and (3) the retention fraction curve.

Table 3.3 Particle-size distribution of a wheat flour sample determined by sieving

Opening of sieve [μm]	Weight [g]	Fraction
<26	33.3	0.0868
26–38	39.2	0.1020
38–53	63.1	0.1640
53–75	97.2	0.2542
75–105	96.9	0.2519
105–149	48.4	0.1266
>149	5.5	0.0146
Total	383.6	

Fig. 3.8 Weight-based frequency distribution of particle size of a wheat flour sample. The *dashed line* indicates the number-based frequency distribution shown in Fig. 3.5

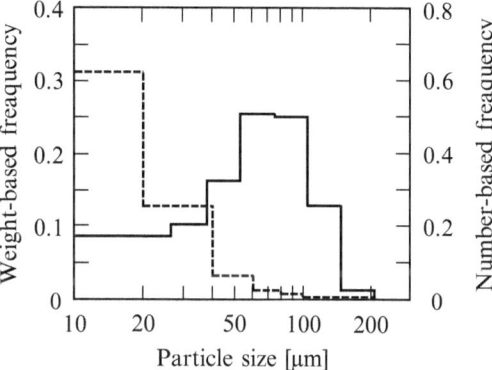

Solution Creating a bar graph by putting particle size on the horizontal axis (logarithmic scale) and weight ratio (frequency) of particles retained on each sieve on the vertical axis gives the solid line in Fig. 3.8. This is the frequency distribution with reference to weight. The frequency distribution based on number of particles (Fig. 3.5b) is represented by the dashed line in Fig. 3.8. When we compare the two lines, it is clear that the distributions differ remarkably depending on the reference. Assuming the true density of wheat flour does not depend on particle size, the number of particles increases inversely with the cube of particle size. For example, if we crush a 200-μm particle into 25-μm particles, we would end up with $(200/25)^3 = 8^3 = 512$ particles. As crushing particles to smaller sizes dramatically increases the number of particles, the number-based frequencies of smaller particles would increase noticeably likewise.

If we sum up the fractions (frequencies) in ascending aperture-size order, we would obtain the cumulative passage fraction at each aperture size (particle size). If we do the summation the other way around in descending aperture-size order, it will give us the cumulative retention fraction. These values are tabulated in Table 3.4. Plotting the cumulative passage and cumulative retention fractions against particle size (on a logarithmic scale) gives the passage fraction and retention fraction curves, respectively, as shown in Fig. 3.9. ◢

Table 3.4 Cumulative passage fraction and cumulative retention fraction of a wheat flour sample

Opening of sieve [μm]	Cumulative passage fraction	Cumulative retention fraction
26	0.087	0.913
38	0.189	0.811
53	0.353	0.647
75	0.607	0.393
105	0.859	0.141
149	0.986	0.015
210	1.000	0.000

Fig. 3.9 Weight-based cumulative passage fraction and cumulative retention fraction of a wheat flour sample

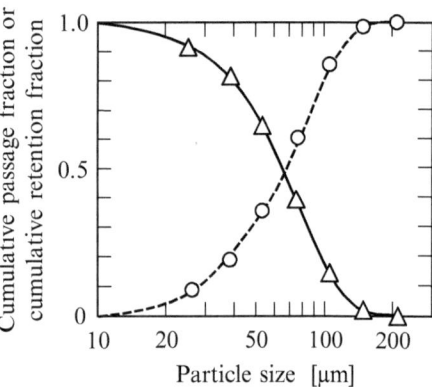

Representing the frequency distribution and cumulative distribution with mathematical expressions may prove to be handy. The log-normal distribution and the *Rosin-Rammler distribution* are among those applied to approximate particle-size distributions. We will cover the Rosin-Rammler distribution which suits well for representing particle-size distributions of ground products like wheat flours. According to the Rosin-Rammler model, cumulative passage fraction, Q, is given by

$$Q = 1 - \exp\left[-(d/d_e)^n\right] \tag{3.4}$$

where d_e is the *granularity characteristic coefficient* which is the particle size when the cumulative passage fraction is 0.632 and n is the *uniformity coefficient* which is a measure of the degree of uniformity in a powdery material in terms of particle size.

Example 3.5 Apply the Rosin-Rammler equation to the distribution of cumulative passage fraction shown in Table 3.4 and determine the granularity characteristic coefficient, d_e, and the uniformity coefficient, n.

Solution Rearranging Q and the exponential term in Eq. (3.4) and taking the natural logarithm of both sides give

$$\ln(1 - Q) = -(d/d_e)^n \tag{3.5}$$

Fig. 3.10 Estimation of the granularity characteristic coefficient, d_e, and the uniformity coefficient, n, of the Rosin-Rammler equation

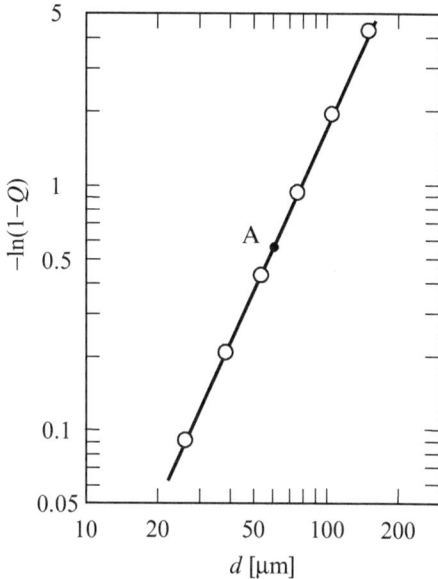

The equation can be transformed by multiplying both sides by -1 and further taking the logarithm of both sides to

$$\log(-\ln(1-Q)) = n \log d - n \log d_e \tag{3.6}$$

Now, plotting $-\ln(1-Q)$ versus particle size, d, on a double logarithmic graph paper would yield a graph like Fig. 3.10. The plot is almost linear, and the uniformity coefficient, n, can be determined to be 2.2 from the slope of the line. Next, substituting the coordinates (60, 0.56) of an arbitrary point (point A in the figure) on the line into the equation gives $\log d_e = (2.2 \log 60 - \log 0.56)/2.2 = 1.89$. Hence, d_e is calculated to be $10^{1.89} = 77.6 \ \mu m$. ◢

3.2.3 Mean Particle Size

An average value (mean particle size) is often used to represent the particle size of a granular material with a wide-spread particle-size distribution. There are several definitions for the average value of particle size, and the typical ones are summarized in Table 3.5. n_i and w_i are the number and mass [kg], respectively, of the particles measuring d_i [m] in diameter. Further, *surface mean diameter* or the Sauter diameter is extensively used for representing the mean particle sizes of droplets and sprayed particles.

Table 3.5 Major computation methods for average particle size

Name	Symbol	Number based	Weight based
Number mean diameter	$d_{1,0}$	$= \Sigma(n_i d_i)/\Sigma n_i$	$= \Sigma(w_i/d_i^2)/\Sigma(w_i/d_i^3)$
Length mean diameter	$d_{2,1}$	$= \Sigma(n_i d_i^2)/\Sigma(n_i d_i)$	$= \Sigma(w_i/d_i)/\Sigma(w_i/d_i^2)$
Surface mean diameter	$d_{3,2}$	$= \Sigma(n_i d_i^3)/\Sigma(n_i d_i^2)$	$= \Sigma w_i/\Sigma(w_i/d_i)$
Volume mean diameter	$d_{4,3}$	$= \Sigma(n_i d_i^4)/\Sigma(n_i d_i^3)$	$= \Sigma(w_i/d_i)/\Sigma w_i$

Example 3.6 Calculate the number mean diameter, $d_{1,0}$, and the surface mean diameter, $d_{3,2}$, from the particle-size distribution based on number of the wheat flour shown in Table 3.2. Use the medians of particle-size ranges as the representative values for each range for the sake of simplicity.

Solution Substituting the values tabulated in Table 3.2 into the equations of number mean diameter, $d_{1,0}$, and surface mean diameter, $d_{3,2}$, stated in Table 3.5 gives

$$d_{1,0} = \frac{331 \cdot 15 + 134 \cdot 30 + 41 \cdot 50 + 15 \cdot 70 + 7 \cdot 90 + 3 \cdot 155}{331 + 134 + 41 + 15 + 7 + 3} = \frac{13180}{531} = 24.8 \ \mu m$$

$$d_{3,2} = \frac{331 \cdot 15^3 + 134 \cdot 30^3 + 41 \cdot 50^3 + 15 \cdot 70^3 + 7 \cdot 90^3 + 3 \cdot 155^3}{331 \cdot 15^2 + 134 \cdot 30^2 + 41 \cdot 50^2 + 15 \cdot 70^2 + 7 \cdot 90^2 + 3 \cdot 155^2}$$

$$= \frac{3.13 \times 10^7}{5.00 \times 10^5} = 62.6 \ \mu m$$

Even from the same set of data, the mean particle size may vary as much as 2.5-folds depending on the definition. We should thus heed the definition of mean diameter when a granular material has a certain particle-size distribution. Moreover, the definition should also be clearly stipulated when presenting a mean diameter. ◢

Example 3.7 Calculate the number mean diameter, $d_{1,0}$, and the volume mean diameter, $d_{4,3}$, of the wheat flour from the weight-based particle-size distribution shown in Table 3.3. Use the medians of particle-size ranges as the representative values for each range as in the case with Example 3.6. Take 10 and 210 μm as the respective minimum and maximum values of the particle-size range.

Solution As described in Example 3.6, by substituting the values presented in Table 3.3 into the equations of number mean diameter, $d_{1,0}$, and volume mean diameter, $d_{4,3}$, in Table 3.5, we obtain the following:

$$d_{1,0} = \frac{33.3/18^2 + 39.2/32^2 + 63.1/45.5^2 + 97.2/64^2 + 96.9/90^2 + 48.4/127^2 + 5.5/179.5^2}{33.3/18^3 + 39.2/32^3 + 63.1/45.5^3 + 97.2/64^3 + 96.9/90^3 + 48.4/127^3 + 5.5/179.5^3}$$

$$= \frac{2.10 \times 10^{-1}}{8.10 \times 10^{-3}} = 26.0 \ \mu m$$

$$d_{4,3} = \frac{33.3 \cdot 18 + 39.2 \cdot 32 + 63.1 \cdot 45.5 + 97.2 \cdot 64 + 96.9 \cdot 90 + 48.4 \cdot 127 + 5.5 \cdot 179.5}{33.3 + 39.2 + 63.1 + 97.2 + 96.9 + 48.4 + 5.5}$$

$$= \frac{2.68 \times 10^4}{383.6} = 69.9 \ \mu m \quad \blacktriangleleft$$

3.3 Texture of Breads

A slice of soft bread dents when pressure is applied on it with a finger, and it slowly rebounds when the pressure is lifted. The described property is known as *elasticity*. Dents are regarded as *deformation*, and the stronger the pushing force is, the bigger the dent would be. The same thing happens during chewing with teeth. The greater the biting force is, the more deformed the bread would be before being eventually cut off. Here, we will discuss the relation between force and deformation while familiarizing you with the way of thinking with the related units of measurement in mind.

3.3.1 Stress-Strain Curve

Exerting the same amount of force on a slice of bread with a finger or with a chopstick will result in different degrees of deformation. Dividing the force applied to a surface by the area over which that force is distributed gives the *stress* value. In the case of a chopstick, the area is smaller than that of a finger and thus the stress is larger. On the other hand, the bread loaf cut into fourths would deform to a greater extent compared to the same loaf being cut into sixths when the same amount of force is applied with a finger. *Strain* is a property of a material, which is obtainable by dividing the amount of deformation by the thickness of the material.

The relation between stress and strain can be measured using a texture analyzer (Fig. 3.11). The amount of deformation, x, is defined as the displacement resulted from the gradual elevation of the stage on which a material (for instance, a slice of bread) to be measured with a thickness of L is placed. The material is put in contact with the plunger with a contact surface area of S, and the repulsion force (internal force) of the material, F, is measured with the load cell. The stress, τ, and strain, γ, are expressed by

$$\tau = F/S \tag{3.7}$$

$$\gamma = x/L \tag{3.8}$$

Stress is equated to the force applied per unit area, and it is expressed in the same units as pressure ($Pa = N/m^2 = (kg \cdot m/s^2)/m^2$), whereas strain is defined as

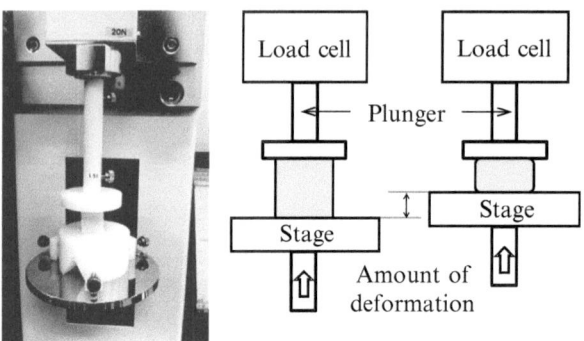

Fig. 3.11 Texture analyzer

Table 3.6 Deformation and force of a bread sample

Deformation [mm]	1.3	2.6	3.9	5.2	6.5	7.8	9.1	10.4	11.05	11.7
Force [N]	0.32	0.68	1.02	1.34	1.81	2.59	4.52	12.01	15.65	13.22
Strain [−]	0.1	0.2	0.3	0.4	0.5	0.6	0.7	0.8	0.85	0.9
Stress [kPa]	0.46	0.97	1.45	1.91	2.58	3.70	6.46	17.16	22.36	18.88

the displacement relative to a reference, which is a *dimensionless quantity* obtained by dividing the deformed length by the reference length (ratio). The graphic representation that illustrates the relation between stress and strain of a particular material is known as its *stress-strain curve*.

Example 3.8 A slice of bread with a thickness of 13 mm is measured for its stress-strain behavior using a texture analyzer, and the results are shown in Table 3.6. The contact area between the sample and the plunger is 7 cm². Construct the stress-strain curve for the sample using the tabulated results.

Solution Strain is estimated by Eq. (3.8) by diving the amount of deformation, x, by the thickness, $L = 13$ mm (=0.013 m), of the bread. Then the corresponding stress, σ, can be computed using Eq. (3.7) by dividing the force exerted on the load cell, F, by the contact area, $S = 7\ \text{cm}^2 = 7 \times 10^{-4}\ \text{m}^2$. The estimates are summarized in the third and fourth rows of Table 3.6. The stress values, τ, occur between 10^2–10^5 Pa $(= \text{N/m}^2)$ and are therefore presented in the kPa using the SI prefix "k" (kilo) representing 10^3. The stress values are now between 0.1 and 10. Constructing a graph of τ against γ on a normal graph paper will yield the stress-strain curve like the one shown in Fig. 3.12. Here, the amount of deformation was targeted at prescribed values such that the strain value increases with 0.1 increments (and 0.05 increments around the region where the stress value peaks). However, a texture analyzer in reality records the amount of deformation and the force applied at far smaller intervals yielding apparently a stress-strain curve. ◢

Fig. 3.12 Stress-strain curve
and Young's modulus of a
bread sample

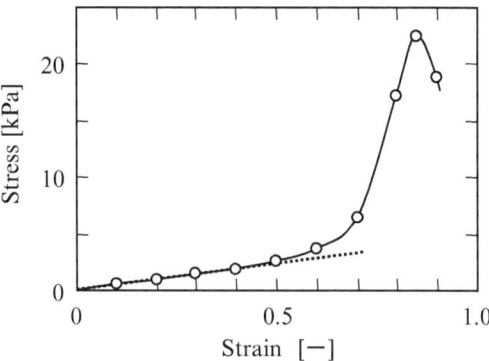

3.3.2 Elastic Solids and Hooke's Law

The way a slice of bread dents being pressed with a finger and rebounds when the finger is lifted resembles that of a spring. *Hooke's law* states that the force in play that displaces a spring by a distance is proportional to the resulted displacement. When a piece of bread or a fish sausage (kamaboko) is pressed within the elastic limit, a similar relation between stress, τ, and strain, γ, (Hooke's law) is established:

$$\tau = E\gamma \tag{3.9}$$

where the elastic constant, E, is regarded as *Young's modulus* or the longitudinal elastic modulus. As mentioned earlier, a spring is an elastic solid in which the force needed to cause a displacement is proportional to the displacement distance. Besides, the relation between stress and strain is expressed by Eq. (3.9). Based on the definitions of stress, τ, and strain, γ, (Eqs. 3.7 and 3.8), Eq. (3.9) can be expressed by

$$F = \frac{ES}{L}x \tag{3.10}$$

The proportionality relation between force, F, and displacement, x, is established. However, in Eq. (3.10), even for the exact same bread, the elastic constant, E, relies on the size, L, of the sample and the contact surface area, S. The elastic constant (Young's modulus) in Eq. (3.9), on the contrary, expresses the specific characteristic of a sample independent of sample size and contact surface area.

Example 3.9 Stress and strain can be deemed proportional within the range of small strains as shown by the stress-strain curve in Fig. 3.12. Find out the Young's modulus for the bread loaf within the aforementioned range.

Solution Drawing a straight line that visually passes the origin and other data points in the vicinity of the origin yields the straight dashed line as shown in

Fig. 3.12. The Young's modulus, E, is determined to be 4.5 kPa from the slope of the dashed line. Whereas by estimating the slope (differential coefficient) at the origin by graphical differentiation (refer to Appendix A2), E is estimated to be 4.6 kPa. These estimated values may also vary depending on the person who does the estimation. Meanwhile, determination of the slope at the origin by numerical differentiation (refer to Appendix A2) using three data points comprising the first two data points and the origin gives

$$E = \frac{(-4)\,(0) + (4)(0.46) - (0.97)}{(2)(0.1)} = 4.35 \text{ kPa}$$

Although the estimates given by numerical differentiation are consistent, the above-described intuitive graphical differentiation often provides values that vary among individuals who did the calculation, not to mention the values obtained by further numerical differentiation. Note that the estimates may differ with people when they are obtained through any sort of manipulation of the empirical data. ◢

3.3.3 Rupture Stress and Rupture Energy

The stress-strain curves of food materials recorded on a texture analyzer generally reveal a shape that resembles that of the one displayed in Fig. 3.13a. Stress, τ, is often proportional to strain, γ, (Eq. 3.9) as demonstrated by the straight line of AB within the range where the amount of deformation is small. Nevertheless, as strain grows bigger, the corresponding stress starts to deviate more and more from the proportionality. This is similar to what one experiences when chewing a food material that greater force is required right before chewing it off compared to the beginning of that particular chew. The described sensation is reflected by the slopes (differential values) of a stress-strain curve. As the amount of deformation increases, the stress increases as well until it reaches point C at which the structure of the food material breaks and the stress decreases thereafter. The strain at point C is named the *rupture strain* (breaking strain), and the corresponding stress is called the *rupture stress* (breaking stress). Further, the stress difference between point C and point E reflects the firmness of a food material.

Graphical representation of the relation between deformation and force would resemble Fig. 3.13a as shown in Fig. 3.13b. The area between the curve of A′C′ and the horizontal axis, in other words the integral of the curve of A′C′ from the amount of deformation = A′ to D′, gives the energy required to bite off the particular food material. Let us consider this chewing phenomenon from the aspects of dimension and unit. (Force) × (Length (amount of deformation)) = (Work or energy), and since the dependent (force) and independent variables (deformation) are multiplied in an integration, the units would be $N \times m = J$ (unit of energy). With

Fig. 3.13 (**a**) Stress-strain curve and (**b**) a curve that depicts the relation between force and deformation

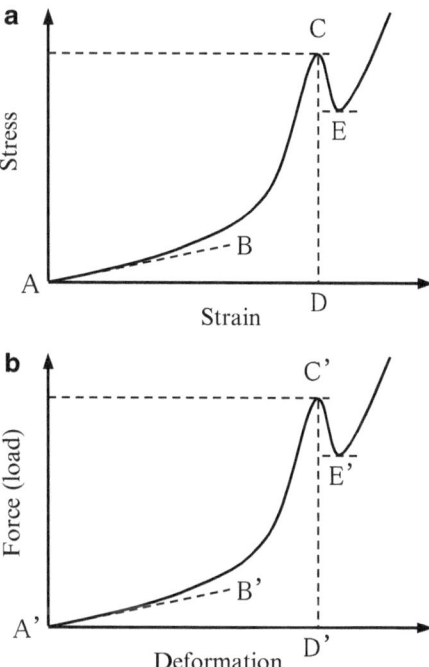

respect to differentiation which is the inverse of integration, the units of the primitive function (dependent variable) are divided by the units of the differentiation variable (independent variable).

Next, considering the integral of the curve of AC from strain $= A$ to strain $= D$ in terms of units for the stress-strain curve in Fig. 3.13a, the units are given by

$$\text{Pa} \times \frac{m}{m} = \frac{N}{m^2} \times \frac{m}{m} = \frac{J}{m^3}$$

which express the energy per unit volume accumulated in the food material in connection with deformation.

Example 3.10 Find out the energy needed to chew off the bread using the data in Table 3.6.

Solution Plotting the graph of force against the amount of deformation using the data gives Fig. 3.14. The shaded area under the curve represents the energy to be obtained. While there are several ways to solve the problem (refer to Appendix A2), we adopt here the trapezoidal rule for numerical integration to find the shaded area, A, by summing up the areas of the trapezoids formed by connecting two consecutive data points on the graph with a straight line:

Fig. 3.14 Energy for chewing a bread

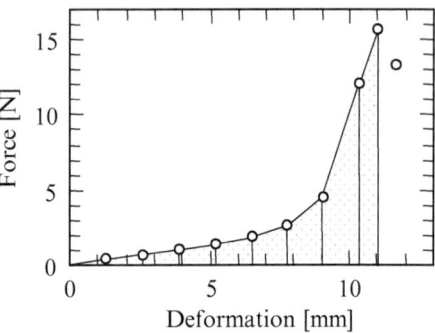

$$A = \frac{1.3}{2}\,(0+0.32) + \frac{1.3}{2}\,(0.32 + 0.68) + \frac{1.3}{2}\,(0.68 + 1.02) + \frac{1.3}{2}\,(1.02 + 1.34)$$

$$+ \frac{1.3}{2}\,(1.34 + 1.81) + \frac{1.3}{2}\,(1.81 + 2.59) + \frac{1.3}{2}\,(2.59 + 4.52)$$

$$+ \frac{1.3}{2}\,(4.52 + 12.01) + \frac{0.65}{2}\,(12.01 + 15.65)$$

$$= (1.3)\left(\frac{0}{2} + 0.32 + 0.68 + 1.02 + 1.34 + 1.81 + 2.59 + 4.52 + \frac{12.01}{2}\right)$$

$$+ \frac{0.65}{2}\,(12.01 + 15.65)$$

$$= 32.76 \approx 32.8$$

Since the force and amount of deformation are in the units of N and mm (millimeter), respectively, the determined energy will hence carry the units of N × mm = mJ (millijoule), where m (milli) is a unit prefix that indicates a thousandth of the unit it precedes. Thus, the energy required to bite off the bread is calculated to be 32.8 mJ = 0.0328 J. ◢

3.4 Spaghetti

We mentioned earlier in the lunch in Chap. 1 that spaghetti is produced from the dough of durum wheat middlings (referred to as durum semolina) kneaded with water and formed into cylindrical shape by extruding the dough through molds with round holes known as dies. Macaroni, linguine etc., collectively regarded as pasta, are also produced by the same process but are just extruded through dies of different shapes. Because fresh pastas produced by the afore-described process spoil rather rapidly, in most cases, they are subjected to drying to reduce the water content before being made available to the market in the form of dried pastas. More specifically, drying a food material to reduce its water content helps improve storage stability and prolong shelf life. Taking pasta as an example, we will further discuss in this subsection the water content and storage stability of food materials and the glass transition phenomena which relates closely with hot air-drying and food storage stability.

3.4.1 Water Activity

The amount of water contained in a food material could be presented either on wet basis or dry basis. Fresh pastas are mixtures of durum semolina and water. The weight of water in a fresh pasta divided by the total weight of the pasta is regarded as the *wet basis moisture content* of the pasta, which have the units of [kg-water/kg-wet material]. On the other hand, the *dry basis moisture content* of the pasta is the ratio of the weight of water to the weight of bone-dry durum semolina, which carries the units of [kg-water/kg-d.m.] or [kg-water/kg-d.s.]. d.m. is the abbreviation for dry matter or dry material and d.s. for dry solid.

Example 3.11 A fresh pasta was prepared from 100 g of durum semolina kneaded with 33 g of water. Determine the wet basis and dry basis moisture contents for the pasta. Fifty grams of the same durum semolina would dry up to the constant weight of 43.5 g in a constant-temperature dryer.

Solution First, determine the wet basis and dry basis moisture contents of the durum semolina. Fifty grams of the durum semolina dries up to the constant weight of 43.5 g which is the weight of dry matter, and thus the weight of water contained is $50 - 43.5 = 6.5$ g. Therefore, the wet basis moisture content is $6.5/50 = 0.13$ g-water/g-wet material $= 0.13$ kg-water/kg-wet material. Meanwhile, the dry basis moisture content is $6.5/43.5 = 0.149$ g-H_2O/g-dry matter $= 0.149$ kg-H_2O/kg-d.m. The difference between the calculated values is due to the different denominators.

The 100 g of durum semolina contains $6.5 \times 100/50 = 13$ g of water and the dry material thus weighs $100 - 13 = 87$ g. Since 33 g of water was added to the dough, the fresh pasta contains $13 + 33 = 46$ g of water, making the wet basis moisture content $46/(100 + 33) = 0.346$ g-water/g-wet material $= 0.346$ kg-water/kg-wet material and the dry basis moisture content $46/87 = 0.529$ g-water/g-d.m. $= 0.529$ kg-water/kg-d.m. Note that when a material contains a significant amount of water, the difference between the wet basis and dry basis moisture contents would vary remarkably. By definition, the wet basis moisture content of a material will never exceed 1 kg-water/kg-wet material (100 % if presented in percentage), but the dry basis moisture content can often go above 100 %. It is not uncommon especially for food materials that contain a considerable amount of water (for instance, boiled pasta) to have a dry basis moisture content of 200 % or 300 %. ◢

Although bread loaves and jams both contain approximately 40 % moisture (0.4 kg-H_2O/kg-wet material), a bread loaf but not jam will grow mold being left at ambient conditions. More specifically, even at similar wet basis (or dry basis) moisture contents, water may take different forms in the food it is present. Therefore, it is inadequate to assess the water contained in a food material merely by moisture content, rendering a separate indicator necessary for evaluation. *Water activity*, a_w, makes the most widely used index for this purpose. The water activity of a food material at a constant temperature is defined as the relative humidity of storage atmosphere, at which the moisture of the material reaches equilibrium (neither increasing nor decreasing). Water activity indicates the strength of interaction between a food material and the water contained in it. Therefore, the water activity of a food that has stronger interactions with water would be lower compared to those which interact less strongly with water even though their moisture contents are identical. *Relative humidity* (or RH) is defined as the ratio of the partial pressure of water vapor, p, to the saturated vapor pressure of pure water, p_s, at a given temperature and is given by

$$\text{RH} = p/p_s \ \text{ or } \ \text{RH} \ [\%] = p/p_s \times 100 \tag{3.11}$$

The term humidity mentioned in weather forecasts refers also to relative humidity. On the other hand, *absolute humidity*, defined as the total mass of water vapor contained in every kilogram of dry air [kg-water vapor/kg-dry air], is a more common term used in the designs and operations of dryers.

There are several ways by which water activity can be measured. Identifying the relative humidity at which no weight change in a food material in question occurs by storing it in atmospheres with different relative humidities would be the most faithful to the definition and simplest method for estimating the water activity. The use of saturated solutions of salts to create atmospheres with known constant relative humidities in closed vessels would be one of the most convenient and useful ways. Different salts have different interaction affinities with water, and by placing a beaker containing a saturated salt solution (with excess of salt as buffer to take up moisture evaporating from food materials during incubation) in a closed vessel, the relative humidity in the vessel is maintained at a particular constant value upon equilibrium. Table 3.7 summarizes the relative humidity values at 30 °C of various saturated salt solutions (known as the water activity of the saturated salt solutions). Further, the water activity of each saturated salt solution, while varying with temperature, also differs slightly among reference sources.

Example 3.12 The weight changes in an udon flour (wheat flour for Japanese wheat noodles) placed in closed containers, the relative humidities of which are maintained at constant values at 30 °C using various saturated salt solutions, were recorded after being kept for 6 h (Table 3.8). w_0 represents the initial weights of the samples and Δw represents the weight changes. Determine the water activity of the udon flour.

Solution Plotting the weight change ratio, $\Delta w/w_0$, versus the relative humidity (water activity) of the saturated salt solutions yields Fig. 3.15. Then connecting the data points with a smooth curve, the curve is found intersecting the horizontal axis

Table 3.7 Water activity of various saturated salt solutions (30 °C) and amount of moisture sorption, q, of a durum semolina sample

Salt	Water activity	q [g-H$_2$O/g-d.m.]
LiCl	0.113	0.053
CH$_3$COOK	0.216	0.072
MgCl$_2$	0.324	0.087
K$_2$CO$_3$	0.432	0.105
Mg(NO$_3$)$_2$	0.514	0.116
NaBr	0.560	0.131
NaNO$_3$	0.730	
NaCl	0.751	0.167
KCl	0.836	0.189
KNO$_3$	0.923	

Table 3.8 Weight change in an udon flour sample

Relative humidity	0.113	0.216	0.324	0.432	0.514	0.560	0.751	0.836	0.923
$\Delta w/w_0$	−0.051	−0.034	−0.021	−0.012	−0.004	0.001	0.023	0.037	0.052

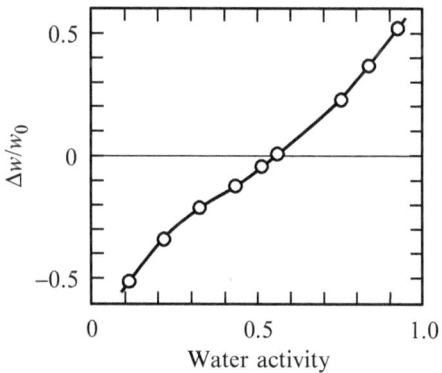

Fig. 3.15 Water activity and weight change

(at which the weight change ratio is zero) at water activity $= 0.552$ which represents the water activity of the flour. ◢

The water activity of a food material closely affects its storage stability. Almost all bacteria are unable to develop at water activity levels below 0.9. However, fungi and yeast can still grow at water activity levels as low as roughly 0.8. Food products that have a water activity adjusted between 0.65 and 0.85 by drying or addition of salt and sugars are called *intermediate-moisture foods* (IMFs). IMFs are semimoist due to their comparatively high moisture contents, but at the same time they also have improved storage stability against microorganisms. Jams, jerked meats, honey, salted fish guts, etc., are typical examples of IMFs. Having all above said, there are still microorganisms such as drought-resistant fungi and osmophilic or osmotolerant yeasts that are able to grow in an environment with water activity levels as low as down to about 0.6. Hence, there is a long list of dried food products whereof water activity is maintained below 0.6 for preservation from microbial decay.

3.4.2 Moisture Sorption Isotherm

As shown in Example 3.12, food materials hold greater amounts of moisture after being stored under higher relative humidity conditions. In other words, food materials take up water (water vapor) from storage atmosphere. The adhering of water molecules to the surface of a material from which a food product is made (including the walls of pores in porous food materials) resulting in higher water concentration on the surface is called *adsorption*. Meanwhile, *absorption* by which water molecules are taken up via chemical or physical action also occurs simultaneously. Because it is hard to distinguish the two, the processes are thus collectively regarded as *sorption*. The curve that displays the relation between the amount of water sorbed by a food material or sorption amount, q [kg-water/kg-dry food material], and water activity, a_w, at a constant temperature is known as its *moisture sorption isotherm* which often bears a resemblance to that shown in Fig. 3.16. The relation is temperature-dependent wherein the amount of moisture sorption generally increases inversely with temperature. Focusing attention on adsorption in the curve shown in Fig. 3.16, in region A water molecules interact with the surface of a solid whereon they adsorb and form a monomolecular layer (monolayer). In region B, adsorption of water molecules continues forming multiple layers on top of the monolayer. In region C, adsorption advances further forming indefinite multilayer of water molecules which the food material has less constraint on but instead are solely held together mechanically.

Moisture sorption isotherms of food materials can usually be described by the *Guggenheim-Anderson-de Boer equation* (Eq. 3.12) which is generally abbreviated as the GAB equation:

Fig. 3.16 Moisture sorption isotherm

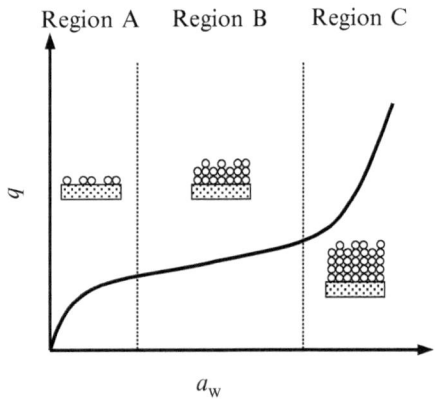

$$q = \frac{q_{m}cKa_{w}}{(1 - Ka_{w})(1 + (c - 1)Ka_{w})} \tag{3.12}$$

where q_{m} is the amount of moisture sorbed upon the formation of a monolayer. Besides, moisture sorption is an exothermic phenomenon, c is the constant related to the difference in the heat of sorption between the monolayer and the layers thereafter, and K is the correction term related to the heat of sorption of water molecules in the second and above layers. The amount of sorbed water, q, is the amount at equilibrium; therefore, the relative humidity of the atmosphere, p/p_{s}, may be used in place of the a_{w} in Eq. (3.12). Estimation of the parameters, q_{m}, c, and K, in Eq. (3.12) from sorption data is exemplified in the following Example 3.13.

Example 3.13 Table 3.7 summarizes the amount of moisture sorption in durum semolina stored in atmospheres with controlled relative humidities using various saturated salt solutions at 30 °C. Construct the moisture sorption isotherm and express the relation with the GAB equation.

Solution Plotting the amount of moisture sorption against the water activity of the saturated salt solutions on a normal graph paper using the data in Table 3.7 gives Fig. 3.17 which resembles Fig. 3.16. Equation (3.12) has three parameters and it is rather tricky to fit them to the experimental data. Since q_{m} is the amount of moisture sorption of monolayer, by connecting the data points in the low water activity region with a smooth curve, the approximate value of q_{m} is estimated to be 0.09 g-water/g-d.m. from the y-axis at the point where the increment of q decreases to the minimum. Besides, K is assumed to take an approximate value of 1 knowing from experience that K ranges roughly from 0.5 to 1. Further, c is estimated to be 10 based on the two estimates determined earlier. Now, using the Solver add-in in Microsoft Excel® (refer to Appendix A4) to fit the GAB equation to the experimental data with the approximate values of q_{m}, K, and c, the constants are determined to be $q_{m} = 0.0879$ g-water/g-d.m., $K = 0.672$, and $c = 14.2$. The solid curve in Fig. 3.17 is the computational curve obtained by plugging these values into Eq. (3.12). ◢

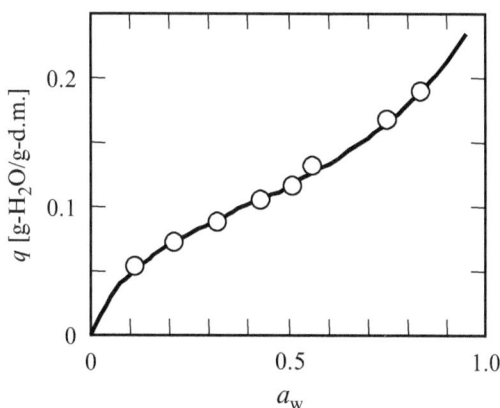

Fig. 3.17 Moisture sorption isotherm of a durum semolina sample

3.4.3   Hot Air-Drying

Drying of food products to reduce moisture content is one of the old methods to improve storage stability with drying in the sun being the oldest of all. However, the method may only apply to small-scale production due to its heavy dependency upon the weather. Whereas in the food industry, *hot air-drying* by which moisture in foods or food ingredients is evaporated and eliminated under a stream of hot air is the most widely employed process. Apart from that, *freeze-drying* by which foods or food ingredients are frozen and then placed under vacuum for sublimation of the frozen water is also practiced in the food industry. Instant coffee makes a good example as a food product that has been made available to the market by these two drying processes. Refer to Chap. 4 for further details.

Dried spaghetti is produced by sun drying in old times, but they are subjected to hot air-drying instead in the production process nowadays. As drying progresses, water is eliminated resulting in the decrease in weight. In practice, fresh spaghetti strands are suspended like string curtains and exposed to a stream of hot air in the drying process. Let us look at the weight change in a single fresh spaghetti strand during the drying process (Fig. 3.18) throughout which the temperature of the strand is also recorded with an ultrafine thermometer embedded in the strand. Drying the strand of fresh spaghetti in a stream of air with constant temperature and relative humidity, the weight and temperature of the strand change with time as represented by the solid and dashed lines, respectively, in Fig. 3.19. At the beginning of the drying process, the weight hardly decreases and the temperature only increases slightly. This particular stage of the process (I in Fig. 3.19) is named the *preheating period*. After that, it enters the stage wherein the temperature stays unchanged, while the weight decreases almost linearly (II in Fig. 3.19; also known as the *constant-rate drying period*). Subsequently the drying process arrives at a stage called the *falling-rate drying period* (III in Fig. 3.19) in which weight loss occurs at slower rates, while the temperature heightens gradually.

The slope of the weight change curve in Fig. 3.19 describes the rate of dehydration. As the mass, m [kg], of the material decreases through the drying

Fig. 3.18 Measurements of weight and temperature

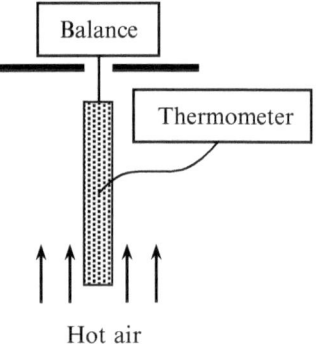

Fig. 3.19 Changes in weight
and temperature

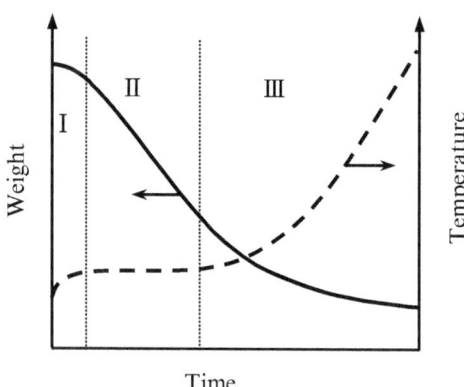

Fig. 3.20 Drying
characteristic curve

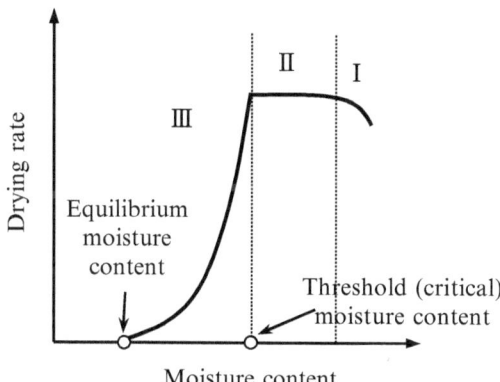

process, the curve has negative slopes, while the drying rates generally take positive values. Therefore, the drying rate, R_w, is given by the slope of the curve multiplied by -1:

$$R_w = -\frac{dm}{dt} \tag{3.13}$$

According to the definition, the afore-described drying rate carries the units of kg-water/s, whereas the drying rate in Sect. 3.4 is defined as the rate of change in moisture content per unit area of drying surface which is presented in the units of (kg-water/kg-d.m.)/(m$^2 \cdot$ s). Figure 3.20 depicts the relation between moisture content and drying rate, and the curve is regarded as a *drying characteristic curve*. The drying rate varies following the curve from right to left as drying progresses. The drying stages marked by "I," "II," and "III" in the figure correspond, respectively, to the preheating, constant-rate drying, and falling-rate drying periods. The moisture content at the transition from constant-rate drying to falling-rate drying is regarded as the *threshold (critical) moisture content*. At one point in time, the moisture content of the drying food material will equilibrate with the

Fig. 3.21 Heat and mass (water) transfers during a drying process

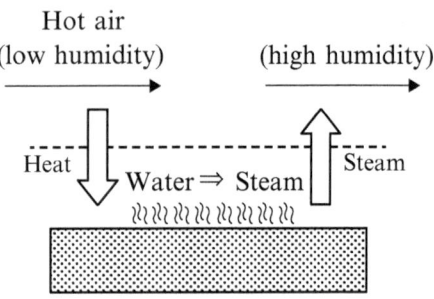

temperature and humidity of the hot air, and the drying rate will drop to 0, indicating the end of the drying process. The moisture content at this point is named the *equilibrium moisture content*. The equilibrium changes with temperature and relative humidity. For instance, elevation of temperature or decrease of relative humidity of the drying medium can further lower the equilibrium moisture content.

During a drying process of a food material, moisture contained in the material is evaporated to the surrounding atmosphere. Heat of vaporization is required for evaporation of moisture to occur, and this heat is afforded by the hot air as a drying medium. A thin film of air (boundary layer) is present adjacent to the surface of a food material (for instance, pasta), which the airflow does not reach. Heat energy is transferred from the hot air stream to the surface of the pasta across this boundary layer (Fig. 3.21). In the constant-rate drying period, the supplied heat is entirely used for evaporation of moisture, and thus, the temperature of the pasta (product temperature) remains constant. Because the water vapor pressure at the surface reaches the saturated water vapor pressure of the product temperature, the water vapor thus migrates into the hot air stream with a comparatively lower vapor pressure (relative humidity). When the moisture content around the surface of the pasta is high, all heat energy supplied by the hot air stream is absorbed for evaporating the moisture. However, as drying advances, the moisture content declines and water molecules deeper down the surface will diffuse to the surface. The distance of diffusion to the surface increases further and the diffusion rate also decreases gradually as drying progresses. At this stage, there comes a time when the supplied heat surpasses the vaporization heat of the contained moisture, resulting in an elevation in product temperature. For this reason, a gradual decrease in drying rate is observed in conjunction with an increase in temperature of drying material during the falling-rate drying period.

3.4.4 Glass Transition

There are two types of solid matters: crystalline and amorphous solids. *Crystalline* solids have regularly and orderly arranged atoms, molecules, or ions, whereas the constituent components in *amorphous* solids do not have regular arrangement. An

Fig. 3.22 Glass transition
curve

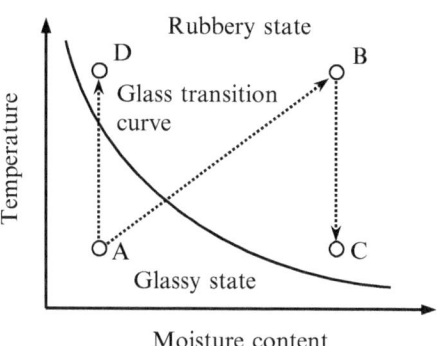

amorphous solid is called a glass; dried pastas, cookies, etc., are food products in the *glassy state*. A dried pasta will snap when it is bent due to its glassy state. On the other hand, when a pasta is allowed to absorb water, it gains flexibility and bends easily. This is known as the *rubbery state* in which the constituent components of the food material are in the molten state. The transition from the glassy to rubbery state and the reverse are regarded as *glass transition*, and the temperature at which the transition occurs is called the *glass transition temperature*, T_g.

T_g exhibits inverse dependency on moisture content; T_g decreases as moisture content increases. The curve that depicts the relation between T_g and moisture content is called a *glass transition curve* (Fig. 3.22). Dried pastas have low moisture contents and are in the glassy state (point A in Fig. 3.22). Cooking a dried pasta in boiling water increases the temperature and returns the moisture content rendering flexibility to the cooked pasta. In other words, it transitions from the glassy to the rubbery state (point B in Fig. 3.22). Even being left to cool down at ambient temperature, the cooked pasta will remain flexible in its rubbery state (point C in Fig. 3.22). Besides, heating a dried pasta in a microwave oven with the pasta being wrapped to prevent moisture loss will also cause the pasta to undergo transition from the glassy to the rubbery state by increasing the temperature (point D in Fig. 3.22), but the reverse occurs, and the pasta hardens when it cools down. A piece of crispy biscuit will likewise transition to the rubbery state if it is being wrapped and heated. Furthermore, if being stored under humid conditions, the piece of biscuit will also absorb moisture leading to elevated moisture content and subsequently transition to the rubbery state.

Glass transition correlates intimately with the storage stability of food products. Food products in the glassy state are very stable because they retain the disordered atomic, molecular, or ionic structures of the liquid state wherein the constituent components are severely immobilized leading to remarkable delay of chemical deterioration. On the contrary, the immobilized components gain mobility in the rubbery state facilitating chemical reactions and rendering the food products more prone to quality deterioration. In summary, it is of pivotal importance to retain food products in the glassy state for long-term storage.

Nonetheless, glass solids are relatively more hygroscopic in comparison to crystalline solids in which the constituents are highly orderly arranged. For example, granulated sugar consisting of crystalline sucrose is nonhygroscopic, whereas castor or superfine sugar consisting of amorphous sucrose takes up moisture more readily. Superfine sugar will sorb moisture on its surface if being left out in a humid atmosphere and undergo transition to the rubbery state. Sucrose in its rubbery state is also in the molten state and the melted particles tend to agglomerate. When the humidity decreases, the agglomerates will undergo transition back to the glassy state. The repeated transitions back and forth between the glassy and rubbery states will cause the fine amorphous sucrose particles to agglomerate and grow in size into large lumps. The described phenomenon can also be observed for instant coffee, milk powder, etc.

Exercise

3.1 Measure the Feret diameter, the Martin diameter, and the Krumbein diameter of the rice flour particle shown in Fig. 3.23.

3.2 Table 3.9 summarizes the particle-size distribution of a jet-milled rice flour. (I) Construct the frequency and cumulative distributions. (II) Assuming that the cumulative distribution can be described by the Rosin-Rammler equation, determine the granularity characteristic coefficient and the uniformity coefficient. And (III) determine the number mean diameter and the surface mean diameter.

3.3 Define the median diameter and the mode diameter when the particle size of a powdery material follows a distribution. Further, find out the median diameter and the mode diameter for the rice flour in Exercise 3.2.

Fig. 3.23 Rice flour particle (*Joshinko* (top-grade rice flour from non-glutinous rice))

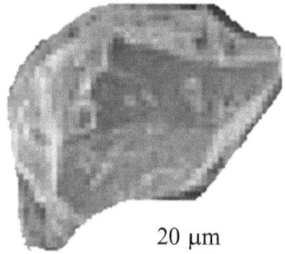

20 μm

Table 3.9 Particle-size distribution of a rice flour sample

Particle diameter [μm]	1.00–1.78	1.78–3.16	3.16–5.62	5.62–10.0
Weight fraction	0.016	0.017	0.028	0.048
Particle diameter [μm]	10.0–17.8	17.8–31.6	31.6–56.2	56.2–100
Weight fraction	0.065	0.232	0.358	0.236

Table 3.10 Deformation and force of a gelatin gel sample

Deformation [mm]	2.5	5.0	7.5	10.0	12.5	15.0	17.5	20.0
Force [N]	4.4	11.4	14.7	20.1	25.7	47.3	80.6	150.3

Table 3.11 Weight change ratio recorded of a dried banana sample

Relative humidity	0.216	0.432	0.56	0.751	0.836
$\Delta w/w_0$	−0.034	−0.016	−0.006	0.011	0.022

Table 3.12 Dry basis moisture content of an udon sample during a drying process

Time [h]	0	1	2	3	4	5
X [kg-H_2O/kg-d.m.]	0.510	0.361	0.296	0.242	0.203	0.178
Time [h]	6	7	8	9	10	
X [kg-H_2O/kg-d.m.]	0.157	0.139	0.122	0.113	0.105	

3.4 The amount of deformation and the force measured of a cylindrical gelatin gel measuring 20 and 30 mm in diameter and height, respectively, using a disk plunger with a larger diameter than the gelatin cylinder are tabulated in Table 3.10. The sample recorded a deformation of 20.3 mm when a force of 162.8 N was applied before it broke. (I) Construct the stress-strain curve. (II) Determine the Young's modulus when the sample is slightly compressed. (III) Find out the breaking strain and the breaking stress. (IV) Compute the amount of energy required to break the gel.

3.5 The weight change ratios, $\Delta w/w_0$, of dried bananas placed over a certain period of time in several closed containers maintained at different constant relative humidities are recorded as shown in Table 3.11. Determine the water activity of the dried bananas.

3.6 A fresh udon (Japanese wheat noodle) was dried in an air stream of 40 °C and 70 % RH, and the change in moisture content of the udon through the drying process was recorded as shown in Table 3.12. (I) Construct the drying curve. (II) Determine the drying rate by graphical differentiation of the drying curve, and then construct the drying characteristic curve. (III) Determine the drying rate by numerical differentiation and construct the drying characteristic curve. (IV) Find out the critical moisture content and the equilibrium moisture content of the udon under this drying condition.

3.7 Elaborate as to how a crispy cookie in the glassy state can be kept from transitioning to the rubbery state during storage. And explain how a cookie that has undergone transition to the rubbery state and became soggy can be returned to the glassy state to restore its crispness.

Chapter 4
Instant Coffee

Abstract This chapter begins with discussions on various operations, i.e., extraction, concentration, and drying, using manufacturing process of instant coffee as an example. Solid-liquid extraction, the same as that employed for obtaining coffee extract from coffee beans, is widely used in food processing. First, we elaborate on how to describe extraction efficiency and extraction rate; we then turn to calculation method of extraction time. Next, we introduce the most widely used equipment in food processing, the evaporative concentrator, and also discuss the calculations involved. Finally, the chapter presents the principles and characteristics of spray-drying and freeze-drying prior to discussing the expression of amount of water vapor contained in air and the design of dryers.

Keywords Extraction • Diffusion • Mass flux • Gurney-Lurie chart • Condensation • Spray-drying • Freeze-drying • Moisture content • Humidity

4.1 Manufacturing Processes of Instant Coffee

Instant coffee encounters are very common in our daily life. The raw material of instant coffee is definitely coffee beans, but what kind of processes has this biomaterial undergone to produce the coffee powders with that pleasant-smelling aroma? In this chapter, we will take instant coffee as an example to discuss the various operations (extraction, concentration, and drying) involved in the manufacturing processes of instant foods which are gaining increasing popularity in this modern age of busy life.

Instant coffee is available in the market in the form of powder produced through three main process operations (Fig. 4.1). Roasted coffee beans are ground to appropriate sizes and then subjected to hot water extraction at 150–180 °C. This operation is similar to brewing drip coffee at home, but the liquid coffee obtained by the industrial extraction operation has higher soluble solid concentrations ranging between 25 and 35 % (w/w). The extraction liquid (coffee extract) is further concentrated to attain even higher concentrations of soluble solids. Finally, water is evaporated off the coffee extract to produce instant coffee in the powder form. As mentioned earlier in Chap. 1, there are two drying methods commonly used for the manufacturing of instant coffee: spray-drying and freeze-drying. Spray-dried coffee powders are

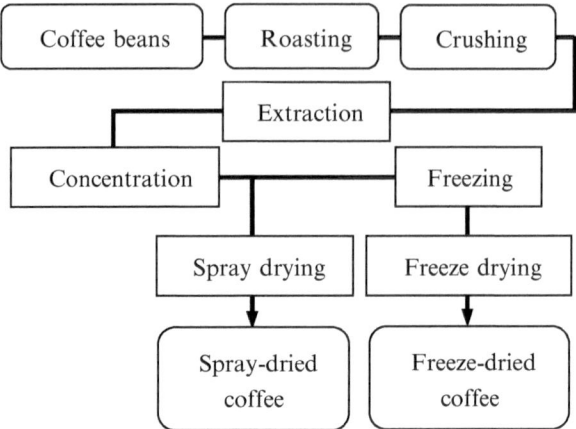

Fig. 4.1 Manufacturing processes of instant coffee powders

relatively finer and more free flowing compared to the freeze-dried counterparts. Spray-drying of coffee extracts to produce instant coffee powders necessitates careful selection of operating conditions due to the exposure of the extracts to hot air at high temperatures which might consequently cause the loss of aroma.

4.2 Extraction

The operation of drawing out the components contained within a solid into a liquid is a common operation in food processing. The coffee extracts from coffee beans and the soybean oil from the soybeans are examples of food ingredients obtained by this separation process called *solid-liquid extraction*. The substances extracted are known as *solutes*, the solids (raw materials) from which the solutes are extracted are called *extraction feeds*, and the media (water and organic solvents) into which the solutes dissolve are regarded as *extractants*.

In the case of food, the extraction feeds are usually cellular cytoplasms as the solutes are comprised in the cells. Target solutes are separated from an extraction feed by a series of mechanisms. When the extraction feed is mixed with an extractant, the extractant will first penetrate into the extraction feed and then solubilize the solutes dispersed inside the solid and finally elute the solutes from the solid. Solid-liquid extraction generally occurs very slowly. Therefore, various measures, e.g., crushing to downsize extraction feed particles and drying to disrupt cell wall, are taken to facilitate the migration of extractant into the solid. In the manufacturing process of instant coffee, the roasted coffee beans are ground into particles of sizes about 1.5 mm. In addition, hot water of high temperatures is used to dissolve the constituent components (carbohydrates) inside the coffee beans because of higher penetration rates of water at elevated temperatures.

Fig. 4.2 Extraction process (migration of solutes and extractant)

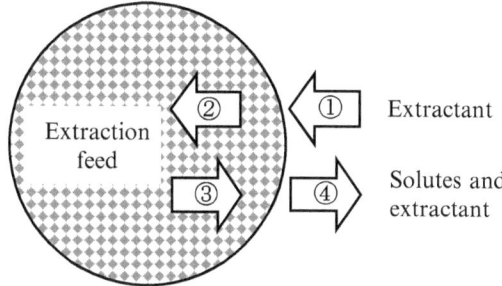

The plausible mechanisms of extraction of solutes from an extraction feed by the use of an extractant are elucidated by the four steps shown in Fig. 4.2. ① The extractant comes into contact with the surface of the extraction feed, ② the extractant penetrates the extraction feed and dissolves the solutes, ③ the extractant migrates back to the surface of the extraction feed carrying the dissolved solutes, ④ and the solutes are eluted into the bulk phase of extractant (main body of extractant).

It is difficult to generally analyze the processes of which the solutes present inside an extraction feed dissolve in an extractant and then migrate toward the surface of the solid. However, the difficulty level of extraction may be assessed under circumstances where certain conditions are met:

(a) When the solutes are present around the surface of the extraction feed forming a layer rich in solutes, extraction may occur rapidly until a near-equilibrium condition is reached.
(b) When the solutes are distributed evenly inside the extraction feed, the migration of solutes within the solid will be the rate-limiting step. In this case, agitation of the whole extraction system would not improve much the extraction rate, but instead it would be more effective to crush the solid to smaller sizes.
(c) When the solutes are contained within a cell, penetration of the cell membrane by the solutes is determined by the osmotic pressure and thus sluggish. In this case, pretreatments such as drying for disruption of the cell membrane would be necessary.

4.2.1 Extraction Operations

There is a vast variety of extraction feeds, and they are normally pretreated into advantageous shapes prior to extraction operation. Although an extraction operation, like other operations, can be a batch, continuous, or semicontinuous (semi-batch) operation, a particular extraction feed has to be something that can be transported continuously to enable a continuous extraction operation. Since the transfer of solid is likely to be one of the sources of troubles, this sort of extraction feeds is usually subjected to semi-batchwise extraction.

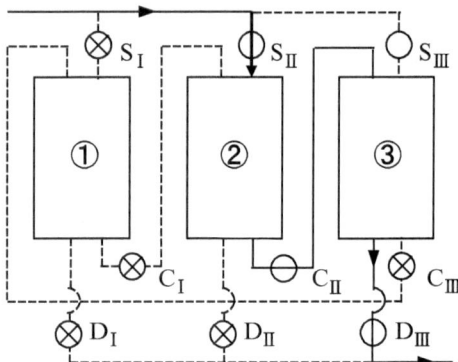

Fig. 4.3 Semicontinuous extractor for solid-liquid extraction. (The solid lines represent flowing extractant and the dashed lines, the intermission of extractant. tower ① (refilling of new feed for extraction (intermitted extraction)) and towers ② and ③ (extraction in progress). In the next step, valves S_{II}, C_{II}, and D_{III} will be closed, and valves S_{III}, C_{III}, and D_I and will be opened; then extraction in tower ② is intermitted for refilling of new feed, while extraction continues to take place in towers ③ and ①)

(1) *Batch extraction*: It is the simplest extraction method of all. In industrial practice, the extraction feed is placed in a container into which the extractant is subsequently added and the mixture is agitated at a certain temperature before being separated into extracted solid and extract. This type of extraction operation is called a *single extraction*. The household preparation of soup stock from dried kelp, bonito, etc., is a typical example of single extraction. When the extraction operation is repeated on the same extracted solid using fresh extractant, the operation is regarded as a *multiple extraction*.

(2) *Semicontinuous extraction*: Figure 4.3 shows an example of a semicontinuous extraction operation using three interconnected extraction tanks. The valves are operated such that the fresh extractant passes through the final extracted feed, while the extractant with the highest concentration of solutes passes through the newly refilled extraction feed. The industrial extraction of coffee basically adopts this extraction method.

4.2.2 Extraction Efficiency

In the operation of extracting compounds from a solid using a liquid extractant, the extractant cannot be completely separated from the extracted solid after extraction, so it will inevitably contain a certain amount of extractant. However, the quantity of extractant residue may vary according to the type of separation process, and thus data need to be collected under the actual conditions for analysis of the process. We will look further at extraction in terms of extraction efficiency in this subsection.

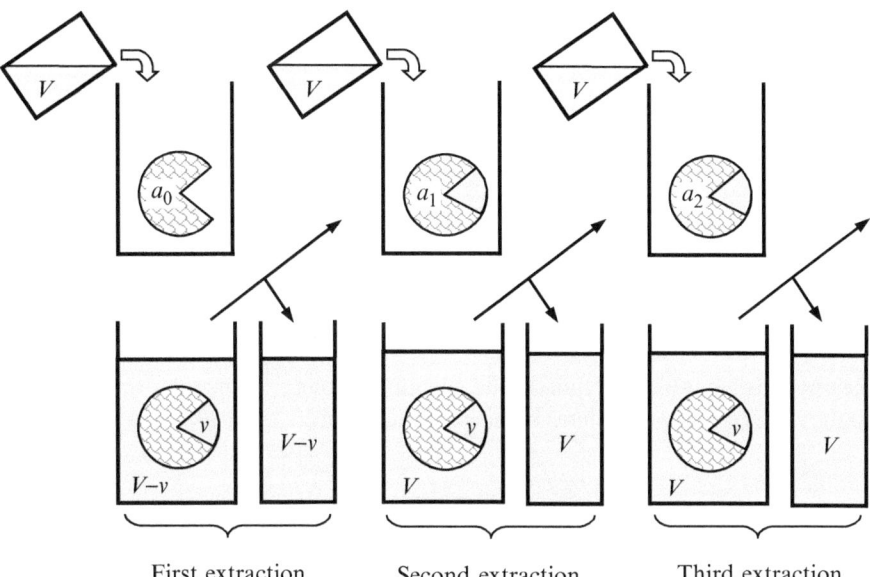

First extraction   Second extraction   Third extraction

Fig. 4.4 Batch extraction

We will make the below assumptions to simplify the calculation (Fig. 4.4). ① The solute concentration in the liquid that remains in the extracted solid (residual liquid) is identical to that of the extract. ② The quantity of extractant in the residual liquid is constant irrespective of solute concentration. ③ Only the solutes in the extraction feed dissolve into the extractant. Let the quantity of extractant used in every extraction operation be V and the quantity of extractant remains in the extracted solid after every extraction operation be v. Let the initial quantity of solutes in the extraction feed be a_0 and the quantities of solutes that remain in the extracted solid after the first, second, $\ldots$, nth extraction operations be $a_1, a_2, \ldots,$ a_n, respectively. After the first extraction operation, a_1 of solute remains dissolved in v of extractant within the extracted solid, and the solute concentration is thus given by a_1/v. According to assumption ①, the extract obtained from the first extraction operation has the same solute concentration; the quantity of solutes in the first extract is hence given by $(a_1/v)(V-v)$. Because the initial quantity of solutes is equal to the sum of solute quantities in both the first extracted solid and the first extract, the mass balance of solutes can be expressed as

$$(a_1/v)\,(V-v) + a_1 = a_0 \qquad (4.1)$$

Note that the extracted solid contains v of residual liquid from the second extraction operation and therefore the same amount of extract as the added extractant, V, will be obtained. Similarly, the mass balances of solutes in the second to nth extracted solids and the second to nth extracts are given by

$$(a_2/v) V + a_2 = a_1$$
$$\cdots \tag{4.2}$$
$$(a_2/v) V + a_n = a_{n-1}$$

Let $r = V/v$ and rewriting the above equations:

$$\left. \begin{aligned} a_1 r &= a_0 \\ a_2 (r + 1) &= a_1 \\ &\cdots \\ a_n (r + 1) &= a_{n-1} \end{aligned} \right\} \tag{4.3}$$

Rearranging Eq. (4.3), the ratio of the quantity of solutes in the nth extracted solid to the initial quantity of solutes in the extraction solid, a_n/a_0, is given by

$$\frac{a_n}{a_0} = \frac{1}{r(r + 1)^{n-1}} \tag{4.4}$$

Let *extraction efficiency* (yield) be E and since $E = 1 - (a_n/a_0)$, therefore

$$1 - E = \frac{1}{r(r + 1)^{n-1}} \tag{4.5}$$

Taking the logarithm of both sides of Eq. (4.5) and rearranging:

$$n = 1 - \frac{\log [r (1 - E)]}{\log (r + 1)} \tag{4.6}$$

Equation (4.6) describes the number of times the extraction operation needs to be performed to achieve the extraction efficiency of E.

Example 4.1 If the quantity ratio of the extractant used in every extraction operation to the extractant remains in the extracted solid, r is equal to 5 for a particular extraction operation, what is the extraction efficiency, E, when the operation is repeated once?

Solution Substituting $r = 5$ and $n = 2$ into Eq. (4.5),

$$1 - E = \frac{1}{5(5 + 1)^{2-1}} = \frac{1}{(5)(6)} = 0.033$$

Therefore, the extraction efficiency is $E = 1 - 0.033 = 0.967 = 96.7\%$. ◢

Example 4.2 If the total amount of extractant used in the two extraction operations in Example 4.1 is to be used for one single extraction operation, what would the extraction efficiency, E, then be? On the other hand, if the amount of extractant used in one extraction operation in Example 4.1 is to be halved, how many times has the extraction operation to be performed in order to attain that efficiency achieved in Example 4.1?

Solution For a single extraction with twice the amount of extractant, $r = 10$ and $n = 1$ are plugged into Eq. (4.4) for calculating a_n/a_0. The extraction efficiency, E, is then determined to be $E = 1 - 0.1 = 0.9 = 90\%$, indicating that the two-time extraction gives a higher yield. Meanwhile, in the case where the extractant quantity is halved, by substituting $r = 2.5$ and $E = 0.967$ in Eq. (4.6), $n = 1 - \log[2.5(1 - 0.967)] / \log(2.5 + 1) = 2.99$, indicating that the extraction operation has to be carried out 3 times. ◢

4.2.3 Extraction Rate

(1) *Diffusion and mass transfer*: Basic knowledge about diffusion is essential when it comes to understanding the relationship between extraction time and efficiency in an extraction operation. When a droplet of a dye is dripped into water in a container, only the water in the vicinity of the dye droplet is colored at first. However, the dye spreads out all over the container as time passes until the water is evenly colored. The dye disperses and migrates even in still water due to thermal motion of the water molecules as well as of the dye molecules themselves. Thermal motion of molecules is random in origin, and the movement of molecules resulting from this motion is referred to as *diffusion (molecular diffusion)*. Like the transfer of heat from regions of high temperature to regions of low temperature, the diffusional movement of a substance occurs in the direction from regions of high concentration toward those of low concentration. As illustrated in Fig. 4.5, the concentration of substance A at point B, C_B, is relatively higher than the concentration of A at point C, C_C; thus A moves in the direction from point B to point C. Let the distance between points B and C be L and the coordinate axis from point B toward point C be x; the number of moles (amount of substance) of substance A that diffuses and passes through a unit area normal to the x-axis per unit time, N_A [mol/(m$^2 \cdot$ s)], is called the *mass flux* that according to *Fick's laws* can be expressed as

$$N_A = -D\frac{dC}{dx} \tag{4.7}$$

where D is a characteristic value of substance A known as the *diffusion coefficient* [m^2/s] and C is the concentration of A. dC/dx denotes the slope of concentration (concentration gradient) of A in the direction of x, which apparently has a negative value as shown in Fig. 4.5. The negative (minus) sign

Fig. 4.5 Diffusion through
a slab

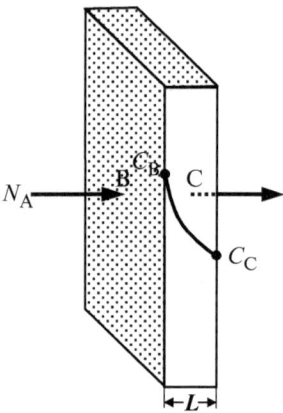

Fig. 4.6 Definition of mass
transfer coefficient

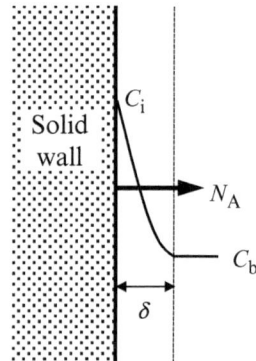

on the right side of Eq. (4.7) is the result of taking the positive direction on the
x-axis as the direction of movement of N_A. When D is constant, integrating Eq.
(4.7) with respect to x from point B to point C gives

$$N_A = D\left(C_B - C_C\right)/L \tag{4.8}$$

The equation shows that N_A is directly proportional to the diffusion coefficient,
D, and the concentration difference, $(C_B - C_C)$, while being inversely proportional to the distance, L. When lumps of sugar are being left as they are in a cup
of coffee, they dissolve sluggishly, but they would dissolve rather quickly when
the coffee is stirred with a spoon. This phenomenon results from the formation
of a *concentration boundary layer* (simply also known as *boundary layer*) as
illustrated in Fig. 4.6 near the interface between the solid of sugar and the
liquid of coffee, where liquid movement hardly occurs and the dissolved sugar
molecules move around by diffusion. Letting the thickness of the boundary
layer be δ [m], the mass flux of sugar molecules inside the layer, N_A can be
described by

$$N_A = D \, (C_i - C_b) \, / \delta \qquad (4.9)$$

where C_i and C_b are the sugar concentrations at the sugar (solid)-liquid interface and in the bulk of the liquid (the liquid region distant from the solid-liquid interface). By stirring the liquid, the boundary layer thins down (δ decreases), and N_A increases; therefore the dissolution of sugar is quickened. Since δ is heavily dependent on the flow condition of the liquid near the solid surface, the term D/δ in Eq. (4.9) may be represented en bloc as the *mass transfer coefficient*, k_m, in the units of [m/s]:

$$N_A = k_m \, (C_i - C_b) \qquad (4.10)$$

The mass transfer coefficient, k_m, varies with the flow condition at the surface of a solid and the shape of the solid and is measured experimentally.

(2) *Calculation of extraction time using the Gurney-Lurie charts*: In an extraction process, the solute concentration in the extraction feed changes with time and region. Estimation of equipment size and extraction time would require analyses that take into consideration the rate at which diffusion of a solute occurs from an extraction feed to the bulk of extractant. These analyses demand the knowledge of partial differentiation equations, and the calculation of the diffusion process is rather complicated that it falls outside the scope of this book. We will discuss here the calculation methods for determining the relationship between solute concentration and extraction time at a certain location inside an extraction feed using the Gurney-Lurie charts for cases wherein the extraction feeds have relatively simple shapes like slab, cylinder, and sphere.

Imagine a process by which the solute contained in a spherical solid of radius R that sits in an extractant is being extracted. The following three assumptions are made in order to simplify the calculation. ① Penetration of the extractant takes place rapidly and the amount of extractant within the solid remains constant at all times. ② The amount of extractant is sufficiently larger than that of the solute in the solid such that the solute concentration in the bulk of extractant is always constant (typically 0). ③ The solute extracted from the solid migrates from the surface of solid by diffusion into the bulk of extractant according to Eq. (4.10). Let the initial solute concentration in the extraction feed be C_0 [kg-solute/kg-solid]. After time t has elapsed since the commencement of the extraction operation, the relationship between the solute concentration, C, and time, t, at the point located r away from the center of the solid can be described by the relationship of the dimensionless concentration, $Y = (C_1 - C)/(C_1 - C_0)$ vs. Dt/R^2 as shown in the chart in Fig. 4.7a, where D is the diffusion coefficient of solute inside the solid (extraction feed) and C_1 is the solute concentration at the surface of the solid. The chart is presented for two different parameters of $m = D/(k_m \cdot R)$ and $n = r/R$, where k_m is the mass transfer coefficient in the extraction feed in equilibrium with that in the bulk of extractant [kg-solute/kg-solid]. When the extractant at the boundary layer

Fig. 4.7 The Gurney-Lurie charts for unsteady mass transfer in (**a**) sphere, (**b**) slab, and (**c**) cylinder

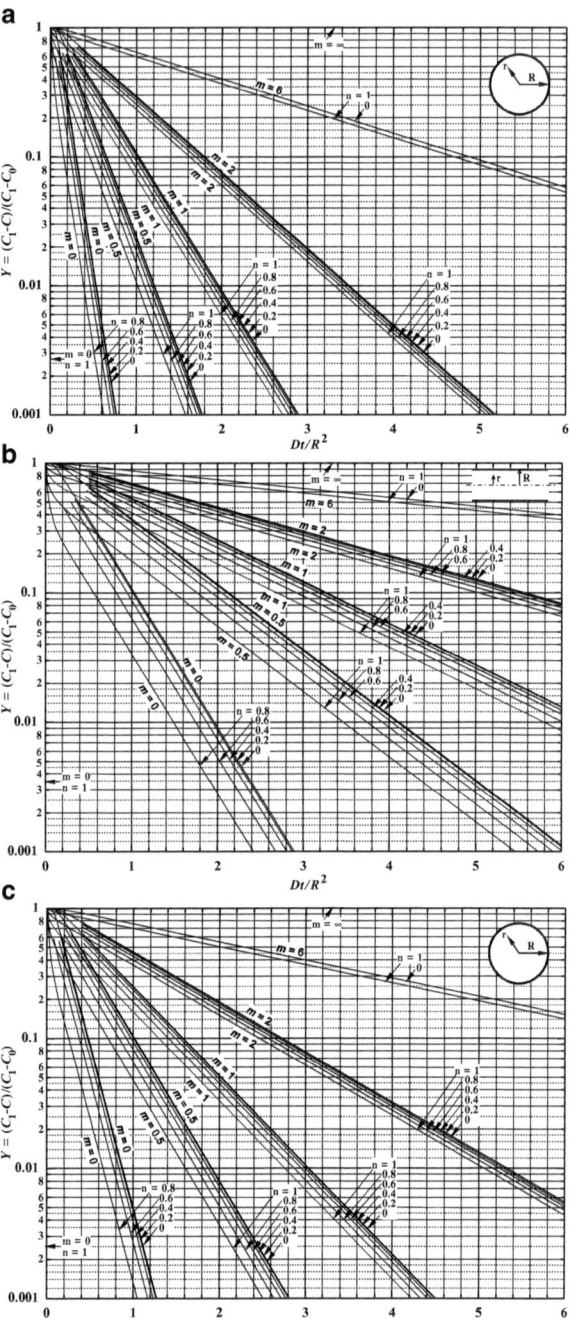

around the solid particles is well stirred, $k_m R$ may be presumed to be sufficiently greater than D, and thus $m = 0$. In addition, the chart for $n = 0$ may be used to determine the solute concentration at the center of the particle (extraction feed) ($r = 0$). As shown in Fig. 4.7b, c, charts analogous to Fig. 4.7a are available for extraction of solute from the upper and lower surfaces of an infinitely long slab of thickness $2R$ and from the side surface of an infinitely long cylinder of radius R. The charts depicted in Fig. 4.7a–c are known as the Gurney-Lurie charts, the applications of which are explained in the succeeding examples.

Example 4.3 The vegetable oil contained in a spherical solid particle with a diameter of 10 mm is being extracted in hexane. If the initial oil concentration in the particle, C_0, is 0.5 kg-solute/kg-solid, determine the oil concentration at the center of the particle after 20 min of extraction in the cases where (a) the mass transfer coefficient around the particle, k_m, approaches infinity and (b) $k_m = 4 \times 10^{-6}$ m/s. Let the diffusion coefficient of oil in the particle, D, be 1×10^{-8} m²/s and the oil concentration in hexane be 0.

Solution (a) $Dt/R^2 = (10^{-8})(20 \times 60)/0.005^2 = 0.48$. Drawing a vertical line across the x-axis at 0.48 and reading the y-coordinate of the intersection of this line with the straight line of $n = 0$ among those in the group of $m = 0$, $Y = (C_1 - C)/(C_1 - C_0)$ is determined to be 0.018. Because $C_1 = 0$, $C = (0.018) (0.5) = 0.009$ kg-solute/kg-solid.

(b) $m = D/(k_m \cdot R) = (10^{-8})/(4 \times 10^{-6} \times 0.005) = 0.5$. Reading the y-coordinate of the intersection of the straight line of $n = 0$ among those in the group of $m = 0.5$ and the vertical straight line at $Dt/R^2 = 0.48$, Y is determined to be 0.21. Therefore, $C = 0.105$ kg-solute/kg-solid which is approximately tenfold of that for the (a) case. As presented in this example, if the concentration boundary layer creates a resistance to mass transfer, the extraction rate will drop, and greater quantities of solute will remain in the extraction feed with the same length of extraction time. ◢

More often in practical extraction operations, we will need to compute the average solute concentration of the entire extraction feed instead of that at specific locations inside it. The Gurney-Lurie charts and the analogous graphical representations discussed previously may prove to be handy for determining the relationship between the average solute concentration and time. Letting the extractant be well stirred such that m is approximated to 0, the solute concentration at the surface of the extraction feed, C_s, is thus constant. At this point, the relationships between Y ($= (C_s - C_{av})/(C_s - C_0)$; the ratio of the concentration difference between C_s and the average solute concentration, C_{av}, to the concentration difference between C_s and C_0) and extraction time, t, for cuboidal, cylindrical, and spherical extraction feeds are presented in Fig. 4.8 (called the average concentration Gurney-Lurie charts). D on the horizontal axis denotes the diffusion coefficient of solute within the extraction feed. While Y_a, Y_b, and Y_c on the vertical axis denote the Y values when solutes are extracted from planes A, B, and C of a cuboid, respectively, Y_r and Y_s represent the respective Y values for extraction of solutes through the side surface of a

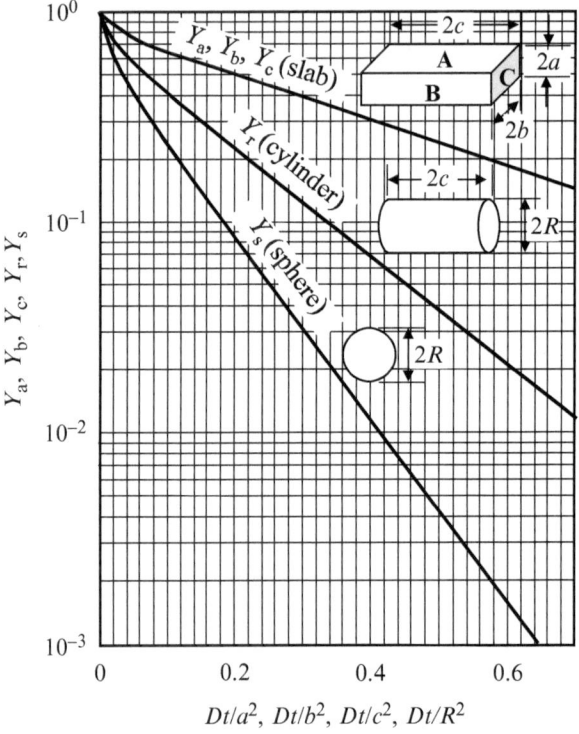

Fig. 4.8 The Gurney-Lurie chart for average concentrations

cylinder and the surface of a sphere. When extraction is performed only through planes A and B, Y is given by

$$Y = Y_a Y_b \qquad (4.11)$$

Whereas for extraction carried out through the three planes of A, B, and C,

$$Y = Y_a Y_b Y_c \qquad (4.12)$$

Likewise, for extraction through the side surface and the two circular surfaces of a cylinder of radius R and length $2c$,

$$Y = Y_c Y_r \qquad (4.13)$$

Example 4.4 A 1.5-mm $\times$ 2-mm $\times$ 5-mm cuboidal extraction feed is placed inside hexane to extract the oil content. The initial solute concentration in the solid, C_0, is 0.7 kg-solute/kg-solid. Determine the fraction of oil remaining in the solid after

15 min of extraction. Let the diffusion coefficient of the oil inside the extraction feed be $D = 4 \times 10^{-10}$ m^2/s.

Solution Let $R = 0.75$ mm, $b = 1$ mm, and $c = 2.5$ mm, thus $Dt/R^2 = (4 \times 10^{-10})$ $(15 \times 60)/(0.00075)^2 = 0.64$, $Dt/b^2 = (4 \times 10^{-10})$ $(15 \times 60)/(0.001)^2 = 0.36$, and $Dt/c^2 = (4 \times 10^{-10})(15 \times 60)/(0.0025)^2 = 0.0576$. Reading from Fig. 4.7 that $Y_a = 0.16$, $Y_b = 0.33$, and $Y_c = 0.77$ and plugging these values in Eq. (4.12), we determine that $Y = (0.16)$ (0.33) $(0.77) = 0.041$. Hence, there are 4.1 % of fat and oil content still remaining in the extraction feed.

4.3 Concentration

Concentration refers to the operation of removing water from an extract to produce a thicker solution, which has wide applications in the food industry. Extracts of coffee, fruit juice, etc., are normally low in solid content, so they are often subjected to concentration processes. Especially for products like instant coffee which the extracts need to be extensively dried into powdery form, it is necessary to concentrate the extracts to increase the solid content in order to prevent loss of flavor through the drying process and for energy-saving purposes as well. While raw liquids are heated with heat sources such as steam to evaporate water in most cases, the concentration operation that involves cooling down raw liquids to crystallize water out (freeze concentration) has also been employed since recent years. Concentration is a heat transfer operation by either method.

4.3.1 Concentration Equipment

Most concentration equipment is of heating type that utilizes steam as a heat source. Figure 4.9a illustrates a fundamental model that concentrates the raw liquid that sits at the bottom of a cylindrical container by heating the liquid with a multitubular heat exchanger and subsequently evaporating water. Liquid foods often contain heat-labile components, and these components are likely to degrade during the concentration process. Further, the solid concentration also increases along with evaporation of water thus elevating the boiling point of the solution. In view of that, concentration operations are often performed under reduced pressures to maintain a low boiling point, creating a greater temperature difference between the boiling point and the heat source (normally steam) in order to evaporate water in a more efficient manner. While the evaporated water vapor is passed through a condenser and discharged as condensate for some models of concentration equipment, there is also the type of evaporator that recycles it as a heat source (Fig. 4.9b). Such type of equipment is called a *multiple-effect evaporator* as opposed to the *single-effect evaporator* that does not reuse the water vapor. Besides, there are also other

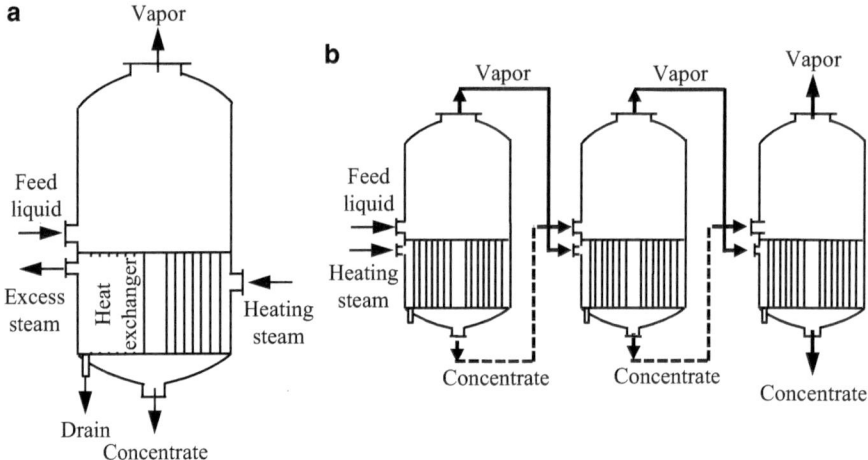

Fig. 4.9 Evaporators. (**a**) Single-effect evaporator. (**b**) Multiple-effect evaporator

types of evaporators, e.g., the thin-film evaporator in which a process liquid to be concentrated flows as a continuous film along inner tube walls or plates while evaporation takes place, the centrifugal thin-film evaporator in which a process liquid is spun into a thin film to facilitate evaporation, etc.

The method of *freeze concentration* was developed for concentrating heat-labile liquid foods. Although the method actually consumes less energy while being capable of producing high-quality concentrates compared to evaporative concentration, the equipment economics has prevented its widespread use. Freeze concentration consists of two basic operations of generation and separation of ice (Fig. 4.10). Ice crystal nuclei are generated using scraped surface heat exchangers (coolers), and the generated fine ice crystals are grown in an agitated vessel, while the process liquid is being circulated through the heat exchanger. In a wash column, the mixture of concentrate and ice is supplied from the bottom of the column, and the ice and wash liquid (concentrate) are separated by a piston mounted with perforated plate that pushes the ice crystals upward. While the packed bed of ice moves slowly upward, wash water is applied at the top of the column to rinse off the liquid adhering to the ice surface and subsequently circulated to mix with the feed liquid.

4.3.2 Design of Concentration Equipment

Let us look at the calculation methods of amount of evaporated water for a single-effect evaporator and also the steam quantity and heat exchange area required for the evaporation. As illustrated in Fig. 4.11, a feed liquid containing components

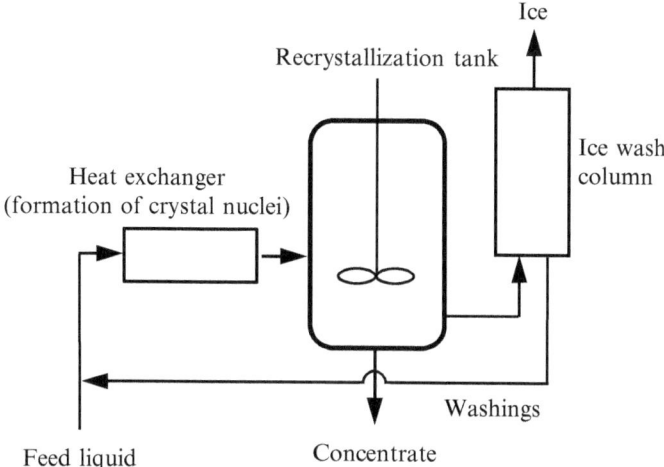

Fig. 4.10 Freeze concentration system

Fig. 4.11 Heat and mass balances for an evaporator

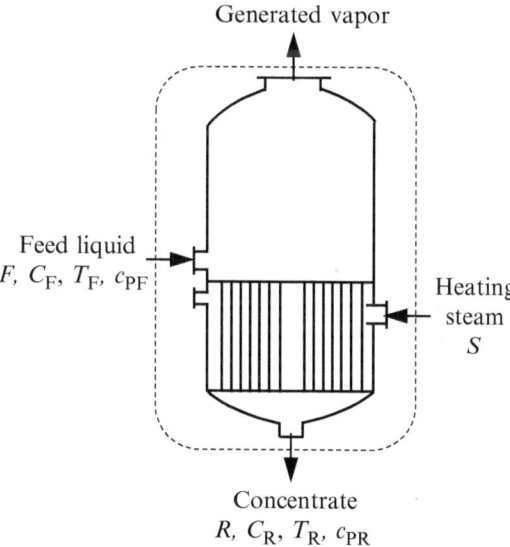

that do not evaporate of concentration C_F [kg/kg-feed] is supplied to a single-effect evaporator at flow rate F [kg/s] and evaporative concentrate. A concentrate of concentration C_R [kg/kg-fluid] is discharged at flow rate R [kg/s]. Let the amount of water vapor generated in the evaporation vessel be V [kg/s]. Considering the mass balance around the evaporator (dashed line-framed area), the overall balance can be expressed as

$$F = R + V \qquad (4.14)$$

And because the non-evaporative components are not present in the water vapor, the balance can therefore be expressed as

$$FC_F = RC_R \tag{4.15}$$

From Eqs. (4.14) and (4.15), the amount of evaporated water (water vapor), V, is given by

$$V = F\left(1 - \frac{C_F}{C_R}\right) \tag{4.16}$$

Next, let us turn to the heat balance. Let the temperature and specific heat capacity of the feed liquid supplied to the equipment be T_F [K] and c_{PF} [J/(kg · K)], respectively, the boiling temperature of concentrate be T_B [K], and the temperature and the specific heat capacity of the concentrate discharged from the evaporator be T_R $(=T_B)$ [K] and c_{PR} [J/(kg · K)], respectively. As depicted in Fig. 4.9a, the process liquid gets heated by absorbing the latent heat of condensation of steam that passes through a vertical shell-and-tube heat exchanger. Letting the latent heat of evaporation of steam be ΔH_S [J/kg], the amount of saturated steam, S [kg/s], needed for heating can then be given by

$$S = \frac{Rc_{PR}(T_R - 273) - Fc_{PF}(T_F - 273) + \Delta H_B V}{\Delta H_S} \tag{4.17}$$

where ΔH_B [J/kg] represents the latent heat of evaporation of the concentrate (latent heat of evaporation of water at T_B).

The area of the heat exchanger used for heating the process liquid (heat transfer area), A [m^2], can be calculated by the following equation from a steady-state heat balance:

$$Q = \Delta H_s S = Rc_{PR}(T_R - 273) - Fc_{PF}(T_F - 273) + \Delta H_B V = UA\Delta T \tag{4.18}$$

where ΔT is the difference between the temperature of steam (heat source) and the boiling point, T_B, of the concentrate and U is the overall heat transfer coefficient in the units of [W/(m^2 · K)].

Example 4.5 An apple juice with a solid concentration of 11 % (w/w) is concentrated to 75 % using a single-effect evaporator. The juice is supplied at 0.7 kg/s, 43 °C. The respective specific heat capacities of the feed liquid and the concentrate are 3.9 kJ/(kg · K) and 2.3 kJ/(kg · K). The temperature of heating steam is 134 °C and its latent heat of evaporation is 2162 kJ/kg. Determine (a) the flow rate, R, of the concentrate, (b) the amount of steam needed for heating, and (c) the required heat transfer area, A, of the heat exchanger when the concentrate has a boiling point of 62 °C (latent heat of evaporation = 2354 kJ/kg). Let the overall heat transfer coefficient of the heat exchanger be 940 W/(m^2 · K).

Solution (a) From Eq. (4.15), $R = FC_F/C_R = (0.7)(0.11/0.75) = 0.103$ kg/s.
(b) S can be computed by using Eq. (4.17). Noting that the terms $T_F - 273$ and $T_R - 273$ in Eq. (4.17) represent the inlet and outlet liquid temperatures in °C, respectively, and given $V = F - R = 0.7 - 0.103 = 0.597$:

$$S = \frac{Rc_{PR}(T_R - 273) - Fc_{PF}(T_F - 273) + \Delta H_B V}{\Delta H_S}$$

$$= \frac{(0.103)(62)(2.3) - (0.7)(43)(3.9) + (2354)(0.597)}{2162} = 0.603 \; kg/s$$

(c) From Eq. (4.18),

$$Q = \Delta H_S S = (0.602)(2162) = 1302 = UA\Delta T = (0.940)(134 - 62)A$$

Solving the equation gives $A = 19.3$ m^2. ◢

4.4 Spray-Drying and Freeze-Drying

Drying coffee concentrates to remove water is the main process along the manufacturing flow of instant coffee powders. Instant coffee can be produced by either *spray-drying* or *freeze-drying* (lyophilization). While spray-drying presents a comparatively more economical means of production, freeze-drying outperforms the former in terms of product quality.

4.4.1 Spray-Drying

Spray-drying is a drying method by which a concentrate is atomized into tiny droplets ranging from several ten to several hundred microns into the stream of hot air, and fine solid particles form as water evaporates quickly from the droplets. A product can be dried very rapidly within 5–30 s compared to other drying methods. Spray-drying requires extremely short time (5–30 s) to dry a product compared to other drying methods, and the product is collected directly in dried powder form. In view of that, this method has been widely employed for turning various liquid foods into dried powders, and recently it has also been gaining attention as a method for preparing nanopowders. While being an indispensable method for manufacturing instant foods in particular coffee, milk powder, seasonings, etc., 8–9 out of 10 powder products containing fragrant components are also produced by spray-drying. Figure 4.12 shows the outline of a spray-dryer. Atomization of process liquids is one of the most critical technologies in spray-drying and is usually performed using a rotary atomizer or a pressure nozzle. Particularly the former is

Fig. 4.12 Spray-dryer
system

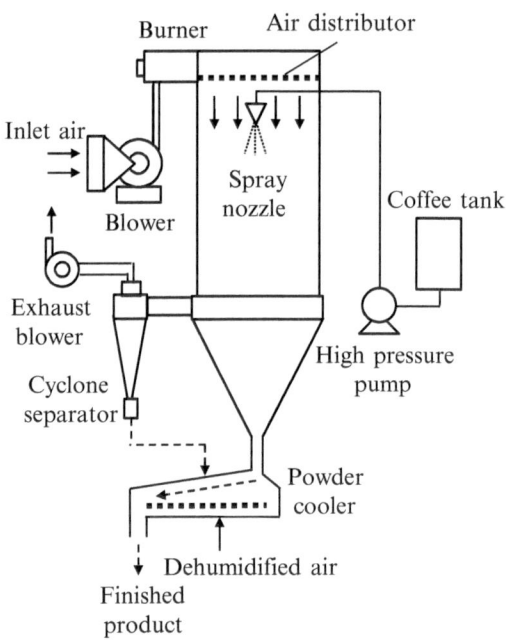

used for atomizing highly viscous liquid foods and slurries containing crystals and
the like due to the possibility to control droplet size by manipulating the rotational
speed. Meanwhile, two-fluid nozzles (pressure nozzles used in conjunction with
compressed air) also demonstrate excellent atomization performance. In a spray-
drying process, atomized droplets are dried in contact with hot air while falling
through the drying chamber from the top. In most cases, air heated with, for
example, a burner is supplied as the drying medium in co-current of the liquid that
is atomized from the top of the drying chamber. The dried powder particles are
discharged from the bottom of the drying chamber and cooled together with the
particles collected in the cyclone (a cylindrical or conical device that vortexes the
mixture of air and powder particles to separate the particles from the spiraling air
stream by the centrifugal force acting on the particles) with a powder cooler.

It is important to obtain a product that retains the flavors and aromas during
the spray-drying of liquid foods such as coffee. Flavors contained in food products
can basically be classified into two categories: water-soluble flavors, e.g., alcohol,
and hydrophobic flavors such as limonene (a type of monocyclic monoterpenoid
found in citrus peels). The principle by which water-soluble flavors are retained
inside the dried particles during a spray-drying process can be conceived as follows.
When a flavor-containing saccharide solution is sprayed into the stream of hot
drying medium, water evaporates rapidly from the droplets forming a film with
a low moisture content (less than 10 % ($= 0.1$ kg-H_2O/kg-d.m.)) at the droplet
surface. The diffusion coefficients of both flavor compounds and water molecules

depend on saccharide concentration: in regions of high saccharide concentrations (regions of low moisture content below 40 %), the diffusion coefficients of flavor compounds become extremely smaller than that of water molecules. Because the dry film formed at the droplet surface is in such a condition that favors the penetration of water molecules over the flavor compounds, most of the flavor compounds are thus retained within the spray droplets. In the spray-drying of hydrophobic flavors, modified starches and emulsifiers such as gum Arabic are used to prepare O/W emulsions of the hydrophobic flavors for stabilizing the flavor compounds in the aqueous solutions to which wall materials such as maltodextrin are added before the mixtures are finally spray-dried.

4.4.2 Freeze-Drying

Freeze-drying or lyophilization refers to the drying method by which raw materials frozen below their freezing points are held under vacuum (reduced pressures) to eliminate moisture through sublimation of the frozen water. The general characteristics of freeze-drying include the following:

(1) The drying materials are less susceptible to physical and chemical alterations, thermal degradation, loss of aromas, and protein denaturation owing to the fact that the raw materials are dried at low temperatures in a frozen state by sublimation of ice.
(2) As moisture is removed from the drying materials in a frozen state, the end products have porous structures and therefore are easily reconstitutable (return to the state before drying).
(3) It is a time-consuming process as drying occurs slowly at low temperatures.
(4) The end products are more expensive compared to those obtained by other drying methods due to higher operating cost, equipment cost, etc.

Although the detail structures of freeze-dryers may differ depending on size and type, a freeze-dryer consists conceptually of three parts (Fig. 4.13):

(1) *Drying chamber*: Cabinet-type drying chamber in most cases and many of the large-scaled units are able to accommodate trays ranging from 50 to 150 m². Depending on the type of drying material, a drying material may either be frozen prior to freeze-drying, or it may also be frozen in the drying chamber by self-evaporation due to the loss of latent heat of evaporation of water. Although operation temperature and pressure also vary with drying material, the process is often performed at approximately 130 Pa between -10 and -30 °C.
(2) *Trap (condenser)*: A refrigerator with a capacity to cool below -40 °C is required in order to capture the generated vapor.
(3) *Evacuation system*: A vacuum pump and evacuation pipelines that are able to maintain the system pressure at about 1 Pa are necessary. A freeze-dryer consumes comparatively more energy than other types of dryers of the same

Fig. 4.13 Freeze-drying
system

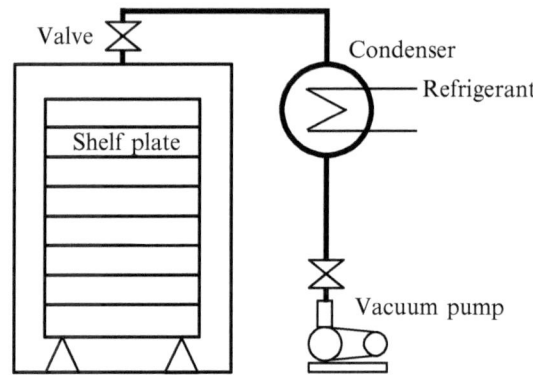

capacity because of the additional energy required for freezing the drying
materials and creating a vacuum environment and also because the sublimation
heat is approximately 13 % greater than the evaporation heat.

The freeze-drying process was developed commercially for drying penicillin.
Its application to food products is rather limited due to high equipment and
operating costs. Recently, however, besides instant coffee, freeze-drying has also
been employed to manufacture relatively inexpensive powdered foods such as
instant miso soup. Although continuous freeze-drying had been challenging, semicontinuous freeze-dryers have recently been developed rendering mass production
possible.

4.4.3   Calculation for Design of Dryer

(1) *Moisture content*: As discussed previously in Sect. 2.2.2, there are two ways to
express the amount of water contained in a food product: *dry basis* and *wet
basis*. A moist material (wet material) turns into a *bone-dry material* when
being completely deprived of moisture. Dry basis moisture content, X, is used
for presenting the amount of water in a wet material by expressing the amount of
water [kg] contained in every kg of the bone-dry material. Because the weight
of bone-dry material does not vary throughout the drying process, it is often
used as the basis for calculation and expression of moisture content as well.
Although moisture content on dry basis is originally dimensionless, it is usually
presented in the units of [kg-water/kg-dry matter] (or [kg-water/kg-d.m.] in
abbreviated form) for the purpose of clarifying its definition. As opposed to
dry basis moisture content, wet basis moisture content, w, expresses the amount
of water [kg] in every kg of the wet material, which is equivalent to the mass
fraction (weight fraction) of water in the wet material. It is simpler to set up
a mass balance on water using dry basis moisture contents, X, for the various
calculations with respect to drying operations as the bases for calculation are

constant before and after drying. The conversion between w and X can be done by

$$X = w/(1 - w) \quad \text{and} \quad w = X/(1 + X) \tag{4.19}$$

From Eq. (4.19), it is apparent that wet basis moisture content, w, can take any positive values below 1, while dry basis moisture content, X, can have values higher than 1 (rather often above 1).

(2) *Equilibrium moisture content*: When a material is being dried with air of constant temperature and humidity, drying stops to progress when the moisture content of the material reaches a certain value. This particular moisture content is known as the *equilibrium moisture content*, X_e, the value of which varies with temperature and humidity of drying air, not to mention the type of food material. The temperature of hot air needs to be increased or the humidity be decreased in order to obtain a finished product with a moisture content lower than the equilibrium moisture content.

(3) *Humidity*: The understanding about the amount of water vapor in air is fundamental in hot air-drying processes. In hot air-drying, water is removed from a drying material into air as water vapor. Due to the limit to the amount of water vapor that air can hold, a drying process may be faced with troubles like insufficient drying and condensation. Although air is generally treated as a mixture of nitrogen and oxygen, it usually carries a few percent of water vapor. Depending on whether water vapor is present in air or not, they are referred to accordingly as moist or dry air. Water vapor concentration in air is basically expressed as partial pressure of water vapor, p_w, but expressions such as humidity are used in general.

Relative humidity (RH), ϕ, is defined as the ratio of the water vapor concentration, ρ_w [kg-water vapor/m^3], in the moist air to the saturated water vapor concentration, ρ_{ws} [kg-water vapor/m^3] at the same temperature. According to the ideal gas law, relative humidity can be expressed as the ratio of partial pressure of water vapor or the ratio of mole fraction of water vapor by

$$\phi = \frac{\rho_w}{\rho_{ws}} = \frac{P_w}{P_{ws}} = \frac{x_w}{x_{ws}} \tag{4.20}$$

where p_w and x_w are the partial pressure and mole fraction of water vapor, respectively, and the letter "s" in the subscript denotes saturation. Relative humidity is generally presented in percentage (%) by multiplying the value calculated by Eq. (4.20), ϕ, by 100.

Since ϕ expresses the relative value of amount of water vapor contained in air, it thus never exceeds 100 % (or 1.0). The humidity mentioned in weather forecasts refers to the relative humidity. As ϕ approaches 100 %, it becomes less likely for water to evaporate into the air; therefore it feels muggy when the relative humidity is high.

Relative humidity is closely related to our daily life and food preservation. However, as per the definition set forth in Eq. (4.20), relative humidity has a downside: the value varies with temperature even though the quantity of water vapor in the air is identical because the saturated water vapor pressure, p_{ws}, changes with air temperature. Relative humidity is not suitable for various calculations in processes that involve temperature changes such as drying, whereas absolute humidity described below would be convenient for calculations of, for instance, the amount of air required for drying. *Absolute humidity*, H [kg-water vapor/kg-dry air], is defined as the mass of water vapor [kg] contained in every kg of dry air. Treating air as an ideal gas, H can be calculated by the following equation:

$$H = \frac{M_w}{M_a} \frac{p_w}{p_t - p_w} = \frac{0.622 p_w}{p_t - p_w} \tag{4.21}$$

where M_w [g/mol] is the molecular mass of water and M_a [g/mol] is that of air, which carry the values of 18.02 and 28.97, respectively. And p_t is the total pressure. The use of the term molecular mass for air is, strictly speaking, inexact because air is a mixture of gases instead of consisting of a single substance. Nonetheless, air is treated as one single component because we are dealing only with moisture in air in drying processes. Summing up the products of the molecular masses of all its major constituent gases and their respective ratios gives 28.97 as the average molecular mass of dry air. In most cases, moist air has a total pressure of 1 atm ($=101.3$ kPa), and hence given the partial pressure of water vapor, p_w, the absolute humidity, H, can be obtained by Eq. (4.21). Similar to relative humidity, absolute humidity likewise is a dimensionless quantity, but it is mostly presented in the fairly self-explanatory units of [kg-water vapor/kg-dry air]. Absolute humidity, H, and relative humidity, ϕ, are interconvertible by the following equations:

$$H = \frac{0.622 \phi p_{ws}}{p_t - \phi p_{ws}} \quad or \quad \phi = \frac{p_t}{p_{ws}} \frac{H}{H + 0.622} \tag{4.22}$$

Example 4.6 What would the relative humidity, ϕ, and absolute humidity, H, of moist air of 25 % RH at 30 °C be when it is cooled down to 10 °C or heated up to 50 °C? The respective saturated water vapor pressures at 10, 30, and 50 °C are 1228, 4243, and 12,334 Pa.

Solution From Eq. (4.20), the water vapor pressure, p_w, of the 30 °C moist air is equal to $4243 \times (25/100) = 1061$ Pa. Hence, also by Eq. (4.20), the RH values at 10 °C and 50 °C would be $\phi_{10} = (1061/1228) \times 100 = 86.4\%$ and $\phi_{50} = (1061/12{,}334) \times 100 = 8.6\%$, respectively. Meanwhile by Eq. (4.21), the absolute humidity, H, is equal to $(0.622) [1061/(101{,}300 - 1061)] = 0.00658$ kg-water vapor/kg-dry air which is a constant value irrespective of temperature. ◢

(4) *Wet-bulb temperature*: Let us suppose a scenario in which a water droplet is present amidst an enormous flow of moist air (at temperature T) with water

evaporating from the droplet surface. If the initial temperature of the water droplet is lower than that of the moist air, heat will transfer from the air to the droplet. While part of the heat is consumed for evaporation of water from the surface as latent heat of evaporation, the remaining is absorbed (as sensible heat) causing a temperature increase in the droplet. As temperature of the droplet continues to increase, water vapor pressure at the droplet surface also increases, and thus more water evaporates. Eventually the system reaches a stage in which all the heat energy transferred from the air to the water droplet is used for evaporation of water, and the temperature of the droplet then stays constant. The temperature of the water droplet at the described stage is referred to as the *wet-bulb temperature*, T_{wb} [K or °C], of the moist air.

(5) *Dew point*: When unsaturated moist air is subjected to cooling, the moist air becomes saturated when the partial pressure of water vapor equals the saturated water vapor pressure. Further cooling below this temperature which is known as the *dew point*, T_d [K or °C], will result in the formation of water droplets by condensation. Vice versa, if the dew point is known, so is the partial pressure of water vapor, p_w, and subsequently ϕ and H can be computed, respectively, by Eqs. (4.20) and (4.21). There are some commercial devices that measure humidity by this principle.

4.4.4 Mass Balance of a Dryer

In the calculations for drying equipment, there are several items that need to be determined apart from the equipment size such as the flow rate of drying medium (hot air), temperatures, and humidity. Calculation of equipment size necessitates determination of the moisture removal rate from a particular process material, which is beyond the scope of this book. Instead, we will turn to the calculation of amount of hot air needed for drying. A process material of dry basis moisture content X_1 [kg-water/kg-d.m.] is supplied at flow rate W [kg/s] to a dryer and dried to dry basis moisture content X_2 (Fig. 4.14). Hot air of humidity H_1 [kg-water vapor/kg-dry air] supplied to the dryer at dry air flow rate G_0 [kg/s] then leaves the dryer with humidity H_2. Let us consider the mass balances of the described drying process at this point in time. First, let us focus on the mass balance on water that enters and leaves the dryer. Since the moisture content was provided on dry basis, it will be simpler to determine also on dry basis the amount of water, V [kg-water/s], to be removed from the process material. The feed rate of process material, W_0 [kg/s], on dry basis into the dryer is given by $W/(1 + X_1)$, the value of which does not vary at the inlet and outlet of the dryer. Hence,

$$V = W_0 (X_1 - X_2) = \frac{W}{1 + X_1} (X_1 - X_2) \tag{4.23}$$

The water evaporated into the hot air increases the humidity at the outlet. Writing the balance on water vapor in the hot air that enters then leaves the dryer, we obtain $G_0 H_1 + V = G_0 H_2$, and the amount of dry air needed for drying can be calculated by

$$G_0 = V/(H_1 - H_2) \tag{4.24}$$

Example 4.7 A coffee extract of 50 % (w/w) solid concentration was spray-dried at the rate of 30 kg/h to produce coffee powder with a dry basis moisture content of 3 %. The humidity, H, of the hot air entering the spray-dryer was 0.01 kg-water vapor/kg-dry air. Determine the flow rate of hot air (as moist air) entering the dryer when the humidity of hot air leaving the dryer equals 0.04 kg-water vapor/kg-dry air.

Solution The dry basis moisture content of the coffee extract is $X_1 = 50/(100 - 50) = 1$ kg-water/kg-d.m. Since the weight of solids is $W_0 = (0.5)(30) = 15$ kg/h, the amount of water to be evaporated is $V = (15)(1 - 0.03) = 14.55$ kg/h. Thus from Eq. (4.24), the quantity of dry air required will be $G_0 = 14.55/(0.04 - 0.01) = 485$ kg-dry air/h which equals $(485)(1+0.01) = 490$ kg/h in terms of moist air. ◢

Exercise

4.1 In an extraction process, an extraction efficiency of 98 % was achieved after the second extraction operation. If the total amount of extractant used for the two extraction operations is to be used for one single extraction operation or to be divided equally into three extraction operations, what will the extraction efficiency be?

4.2 A cheese wheel of 15 cm in diameter and 5 cm in thickness is immersed in a NaCl solution, allowing NaCl to diffuse into the cheese. What would the NaCl concentration be inside the cheese after 5 days of immersion? Assume that the concentration of the NaCl solution is constant throughout the immersion and that at the surface of the cheese wheel immediately reaches an equilibrium concentration of 60 kg/m³ cheese and stays constant thereafter. Let the diffusion coefficient of NaCl inside the cheese be 3×10^{-10} m²/s.

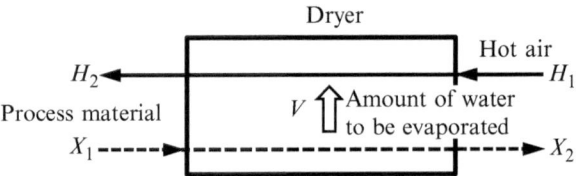

Fig. 4.14 Estimation of air flow rate necessary for drying

4.3 A sugarcane solution of 38 % (w/w) sugar concentration is concentrated using an evaporator at 400 kg/h to 74 % (w/w) sugar concentration. Determine the amount of water evaporated by the evaporator and the amount of concentrate obtained.

4.4 A single-effect evaporator with a heat transfer area of 30 m^2 is used to concentrate a juice of 12 % (w/w) solid concentration to 20 %. The juice enters the evaporator at 50 °C and evaporation occurs at 60 °C. Saturated steam of 100 °C is employed as the heat source. Let the overall heat transfer coefficient of the heat transfer surface (of heat exchanger) be 1000 W/m^2 and the specific heat capacity of the juice be $4.18 - 2.51C$ kJ/(kg · K), where C denotes the solid concentration (mass fraction) of the juice. What are the feed rate of raw juice into the evaporator and the amount of saturated steam required for heating? Let the latent heat of evaporation of water be $2500 - 2.24t$, where t denotes temperature in the unit of °C.

4.5 Thirty thousand cubic meter of air of 70 % RH (at 30 °C) is mixed with 1.0×10^4 m^3 of air of 50 % RH (at 40 °C). Determine the absolute humidity, H, of the mixture air. The respective saturated water vapor pressures at 30 and 40 °C are 4.243 and 7.376 kPa.

4.6 A solid food that contains 15 % (w/w) moisture on wet basis is subjected to hot air-drying by which the moisture content is reduced to 7 %. Part of the outlet hot air is recycled and mixed with the inlet hot air for drying the solid. The humidity of the fresh inlet hot air, the recycled outlet hot air, and the mixture hot air that enters the dryer is 0.01, 0.1, and 0.03 kg-water vapor/kg-dry air, respectively. When the process raw material is fed at 100 kg/h to the dryer, what are the flow rates of the fresh inlet hot air and the recycled outlet hot air, and what is the discharge rate of the finished product?

Chapter 5
Mayonnaise and Margarine

Abstract The margarine spread on toasts, the milk for breakfast and the mayonnaise for dressing the salad for lunch shown in Fig. 1.1 in Chap. 1 are all emulsions. Further, the creamer used in the after-lunch coffee is as well an emulsion. Our daily diet is in fact composed of many foods that are classified as emulsions (or regarded as emulsified foods). The foods like margarine, milk, and mayonnaise discussed in Chap. 1 are typical emulsified foods. Although knowledge concerning emulsion has been burgeoning in recent years along with the advancement in interfacial science, we will learn about the fundamentals of emulsion-forming process holding basically mayonnaise up as an example. Emulsions and liquid food polymers display combined viscous and elastic flow behaviors which are different from those of ordinary fluids (Newtonian fluids) like water. Fluids of this kind are called viscoelastic fluids, the basics of fluid flow of which will be further elaborated in this chapter.

Keywords Emulsion • Dispersed phase • Surfactant • Critical micelle concentration • Gibbs adsorption isotherm • HLB value • Viscosity • Non-Newtonian fluid • Casson's equation • Pseudoplastic fluid • Viscoelastic fluid • Maxwell model • Voigt model • Relaxation time

5.1 Emulsion

5.1.1 Classification of Emulsion

A *dispersion system* refers to a matter system wherein small particles are scattered within another substance in a certain state. The former is known as the *dispersed phase* and the latter the *dispersion medium* (or the *continuous phase*). For instance, in beer where small bubbles are dispersed, the liquid is the dispersion medium, while the air bubbles are the dispersed phase. Meringue made from whisked egg white and whipped cream made from beaten cream are other similar examples.

An *emulsion* is a dispersion system made up of two immiscible liquids, for example, water and oil, of which one of the liquids becomes the dispersion medium while the other is being atomized and scattered in the medium as the dispersed phase (Fig. 5.1). The emulsions that contain tiny oil droplets in an aqueous

© Springer Science+Business Media Singapore 2016

T.L. Neoh et al., *Introduction to Food Manufacturing Engineering*,

DOI 10.1007/978-981-10-0442-1_5

Fig. 5.1 O/W and W/O
emulsions and their examples

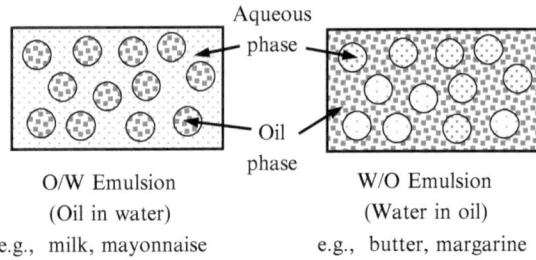

O/W Emulsion   W/O Emulsion
(Oil in water)   (Water in oil)
e.g., milk, mayonnaise   e.g., butter, margarine

continuous medium are called oil-in-water emulsions or generally abbreviated as
O/W emulsions. The milk and mayonnaise illustrated in Fig. 1.1 are good examples
of this type of emulsions. On the other hand, the emulsions wherein water droplets
are being dispersed in oil are named water-in-oil emulsions or *W/O emulsions*. The
margarine is an example of a W/O emulsion. Moreover, there are other types of
emulsions called multiple emulsions where water-in-oil and oil-in-water emulsions
coexist in a single dispersion system. The one wherein even smaller water droplets
are dispersed in the oil droplets of an O/W emulsion is regarded as a water-in-oil-in-
water (W/O/W) emulsion, while the one where still tinier oil droplets are scattered
in the water droplets of an W/O emulsion is referred to as an oil-in-water-in-oil
(O/W/O) emulsion. However, these complex polydispersed systems are rarely found
in food.

5.1.2   Emulsification Operation and Equipment

The operation for preparing an emulsion from two or more immiscible liquids is
called *emulsification*. Emulsification can be achieved by a top-down approach by
which the bulk of the dispersed phase is broken down into smaller droplets in
the dispersion medium and a bottom-up approach by which the dispersed phase
is formed from its constituent substances that are dissolved in the continuous phase,
with the former approach being the major practice in the industry. Mayonnaise can
be made at home by whisking drops of cooking oil (oil phase) into vinegar (water
phase) containing egg yolk (acting as an emulsifier which will be introduced later
in the subsection) to emulsify the oil. This household emulsification method is an
example of the top-down approach.

There are various types of emulsification equipment for industrial use as shown
in Fig. 5.2. A typical homogenizer consists of an external stator and an internal rotor
blade (Fig. 5.2a). The internal rotor blade rotates at high speed in a mixture liquid
of dispersed and continuous phases, spinning and atomizing the dispersed phase
through the windowed or toothed external stator. A colloid mill (Fig. 5.2b) breaks
down the coarse droplets of the dispersed phase in a premixed emulsion by applying
high levels of hydraulic shear and turbulence to the emulsion, while it passes through

Fig. 5.2 Various types of emulsification equipment. (**a**) Homogenizer, (**b**) Colloid mill, (**c**) Ultrasonic emulsification equipment, and (**d**) Higher pressure homogenizer

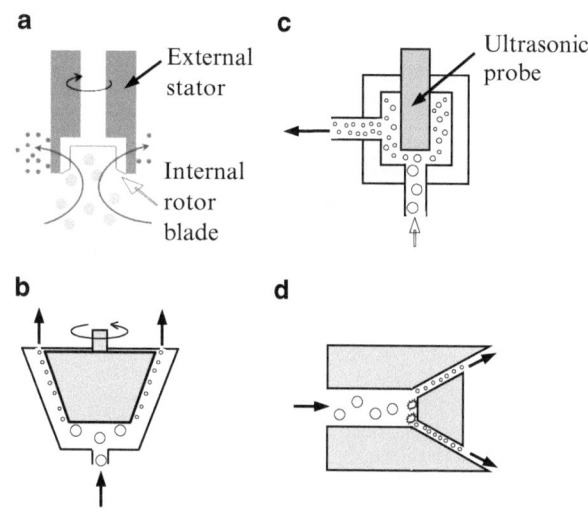

the narrow gap between the stator and the truncated conical rotor that rotates at high speeds. An ultrasonic emulsification equipment (Fig. 5.2c) works by producing ultrasonic vibration in a mixture liquid of dispersed and continuous phases through its ultrasonic probe, generating pressure difference that results in the formation of tiny bubbles (cavitation), thus making a great impact on the dispersed phase and causing it to break down to smaller droplets. In a high-pressure homogenizer (Fig. 5.2d), a liquid of premixed coarse emulsion is forced through a special nozzle (generator) at high velocity by a high-pressure pump, and the emulsion is hence subjected to ultrahigh-speed shear force, shock wave, and cavitation, further disrupting the coarse droplets of dispersed phase.

5.2 Surfactants (Emulsifiers)

5.2.1 Characteristics of Surfactants

On the molecular level, the molecules of a liquid are present in the bulk of the liquid and at the surface of the liquid as illustrated in Fig. 5.3. The molecules in the liquid are surrounded by other molecules and are subject to the intermolecular attractive forces of equal intensity exerted by all adjacent molecules in all directions. The attractive forces between molecules are shared with neighboring molecules causing an energy drop to the molecules. However, in addition to the extremely weak attractive forces exerted by the gaseous molecules above the surface, the surface molecules share the attractive forces with fewer neighboring molecules in the liquid itself, leading to a comparatively smaller drop in energy in the surface molecules.

Fig. 5.3 Forces acting on the atoms, molecules, or ions in the bulk and at the surface of a liquid

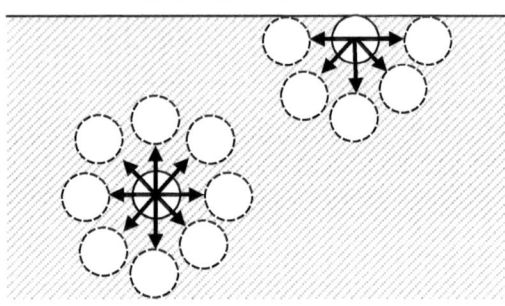

Fig. 5.4 A surfactant molecule

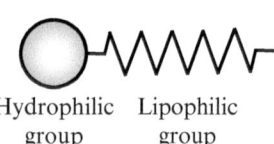

Hydrophilic   Lipophilic
group   group

Hence, the surface molecules have relatively higher energy (surface energy), and they cohere more strongly between each other in order to minimize this energy. This attractive force is known as *surface tension*. Moreover, similar attractive forces known as *interfacial tension* are at work at the interface between two immiscible phases such as oil and water.

Separate-type salad dressings (a type of O/W emulsions) are mixed and shaken vigorously to disperse the oil phase immediately prior to use. The oil and aqueous phases will separate after being left to stand for a short while due to the increase in energy level at the phase boundary resulting from the formation of oil droplets. If surface-active molecules capable of reducing the oil-water interfacial energy are introduced to the dispersion system, the separation process will be impeded. Egg yolks used as one of the ingredients for the preparation of mayonnaise contain a substance called lecithin that exhibits some surfactant properties. Surfactants are predominantly employed for emulsification in the food sector and thus are usually called emulsifiers.

Surfactants (emulsifiers) are amphiphilic substances that contain both hydrophilic regions and hydrophobic regions (Fig. 5.4). When dissolved in water at low concentration, a surfactant is present as monomers (Fig. 5.5). As the concentration increases, although there are still surfactant molecules that dissolve as monomers, the other surfactant molecules start to align in such orientation that their hydrophilic groups stay in the water phase, while the hydrophobic groups stick out into the air phase due to its considerably strong hydrophobicity (nitrogen and oxygen, being the main constituents of air, are scarcely soluble in water). The described arrangement of the surfactant molecules at the surface or interface reduces the surface (interfacial) tension. With further increase in concentration and the surface already covered, the surfactant molecules start aggregating into *micelles* wherein the hydrophobic regions face inward sequestered by the hydrophilic regions that form an outer layer in contact with water. The surfactant concentration above

Fig. 5.5 The relationship between surfactant concentration and surface tension

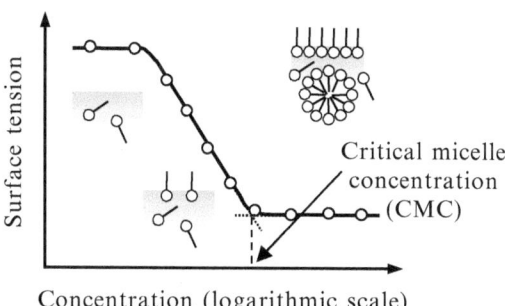

Fig. 5.6 Measurement of surface tension

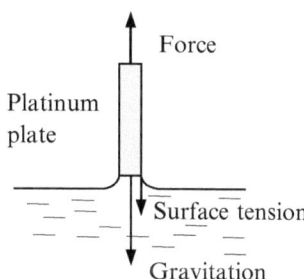

which micelles form is designated as the *critical micelle concentration* (CMC). Although the micelle depicted in Fig. 5.5 is spherical, they are not necessarily spherical. They may be cylindrical or take various other shapes depending on the type of surfactant.

There are several methods for measuring surface tension. The Wilhelmy plate method is one of the most common methods by which surface tension of a liquid is determined from the force required to draw a clean, thin platinum plate immersed in the liquid vertically upward (Fig. 5.6). The relationship between surfactant concentration below CMC, C, and surface tension, γ, is given by the *Gibbs adsorption isotherm*:

$$\Gamma = -\frac{1}{2.30RT} \frac{d\gamma}{d\log C} \tag{5.1}$$

where R [J/(mol·K)] is the gas constant, T [K] is absolute temperature, and Γ [mol/m^2] is the amount of surfactant per unit area of surface regarded as *surface excess* (generally called interfacial excess).

Example 5.1 A surfactant is dissolved in water at prescribed concentrations, and the surface tension of the solutions is measured (at 25 °C) as shown in Table 5.1. Determine the CMC, the surface excess, Γ, and the *area occupied by a single surfactant molecule* at the interface, a. The "m" at the front of the units of surface tension, mN/m, is an SI unit prefix that represents a thousandth of the unit that follows, and the other one at the back indicates meter, a unit of length.

Table 5.1 Concentration of surfactant and surface tension

Concentration, C [mol/L]	4.37×10^{-6}	1.37×10^{-5}	3.43×10^{-5}	6.85×10^{-5}	1.03×10^{-4}
Surface tension, γ [mN/m]	71.8	61.9	47.9	40.2	32.6
Concentration, C [mol/L]	1.37×10^{-4}	3.43×10^{-4}	6.85×10^{-4}	1.03×10^{-3}	
Surface tension, γ [mN/m]	33.2	30.4	29.4	29.1	

Fig. 5.7 Surface excess and critical micelle concentration (CMC)

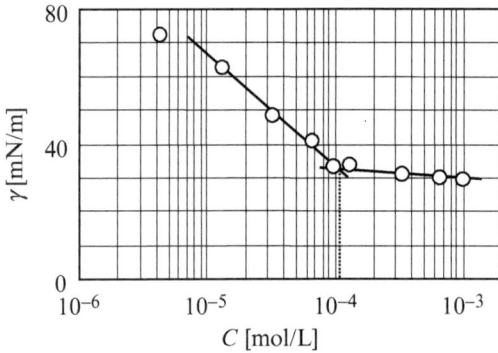

Solution Plotting surface tension, γ, against surfactant concentration, C, on a semilogarithmic graph paper yields Fig. 5.7. The CMC is given by the x coordinate of the intersection point between the downward-sloping and nearly horizontal straight lines to be 1.1×10^{-4} mol/L = 0.11 mmol/L. Besides, the corresponding values of γ when $C = 10^{-4}$ mol/L and 10^{-5} mol/L are 32.8 mN/m = 0.0328 N/m and 66.5 mN/m = 0.0665 N/m, respectively:

$$\frac{d\gamma}{d \log C} = \frac{0.0328 - 0.0665}{\log 10^{-4} - \log 10^{-5}} = \frac{-0.0337}{-4 - (-5)} = -0.0337 \text{N/m}$$

Substituting this equation, $R = 8.31$ J/(mol · K), and $T = 298$ K in Eq. (5.1) gives the surface excess as

$$\Gamma = -\frac{1}{(2.30)(8.31)(298)} (-0.0337) = 5.92 \times 10^{-6} \text{mol/m}^2$$

$J = N \cdot m$; therefore, surface excess carries the units of

$$\frac{1}{(J/(mol \cdot K)) \, K} \frac{N}{m} = \frac{1}{N \cdot m/mol} \frac{N}{m} = \frac{1}{m^2/mol} = \frac{mol}{m^2}$$

which indicates the quantity of surfactant per unit area of the surface [mol/m^2]. Note that a logarithm has no units and thus when logarithmized, the quantities that have units become dimensionless. The number of molecules in one mole of surfactant is given by the Avogadro constant, $\tilde{N} = 6.02 \times 10^{23}$ mol^{-1}. Then we obtain the area occupied by a single surfactant molecule at the interface, a, by taking the inverse of the product of surface excess, Γ, and Avogadro's constant, $\tilde{N}$:

$$a = \frac{1}{\Gamma \tilde{N}} = \frac{1}{(5.92 \times 10^{-6})(6.02 \times 10^{23})} = 2.80 \times 10^{-19} \text{m}^2/\text{molecule}$$

$$= 0.280 \text{ nm}^2/\text{molecule} \; \blacktriangle$$

5.2.2 HLB Value

Surfactants (emulsifiers), resulting from the combination of hydrophilic groups and hydrophobic groups, are vastly diverse in type. The *HLB index* is a measure of the surface-active properties of surfactants. HLB stands for hydrophile-lipophile balance, and it is an index used for the selection of suitable surfactants for intended systems. HLB scale ranges from 0, for highly lipophilic (hydrophobic) surfactants, to 20, for predominantly hydrophilic ones. Figure 5.8 shows the applications of surfactants classified based on HLB scale. Surfactants (emulsifiers) of close properties to the continuous phase are picked for preparation of an emulsion. Hydrophobic emulsifiers of low HLB values are employed to prepare W/O emulsions because of the hydrophobic continuous phase of oil, whereas O/W emulsions are prepared using emulsifiers with high HLB values that are compatible with water as the continuous phase.

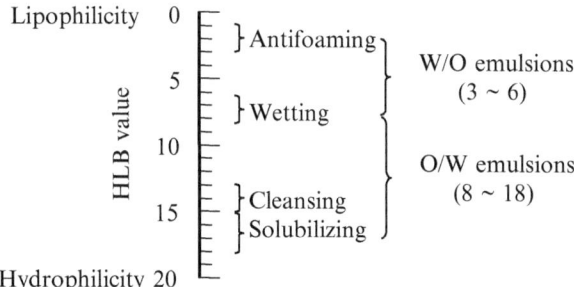

Fig. 5.8 HLB values and applications of surfactants

Fig. 5.9 Dispersion of oil
into water

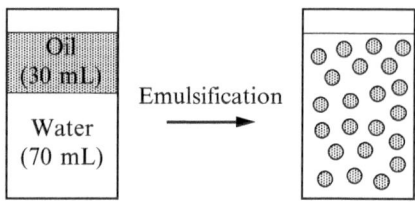

Table 5.2 Increases in interface area and energy due to disruption of oil droplets

Diameter of oil droplet [μm]	Number of droplets [−]	Surface area of an oil droplet [m²]	Interface area [m²]	Energy [J]	Energy (relative value)
Initial state	1	2.0×10^{-3}	2.0×10^{-3}	1.0×10^{-4}	1
10	1.9×10^{11}	3.1×10^{-10}	60	1.2	30,000
1	1.9×10^{14}	3.1×10^{-12}	600	12	300,000
0.1	1.9×10^{17}	3.1×10^{-14}	6000	120	3,000,000

5.2.3   Minimum Energy for Emulsification

As described earlier in this chapter, interfacial tension indicates the surface energy.
Let us look at it in terms of units. Interfacial tension carries the units of N/m, and
multiplying both the denominator and numerator by m (meter, a unit of length) gives

$$\frac{N}{m} \times \frac{m}{m} = \frac{N \cdot m}{m^2} = \frac{J}{m^2}$$

Thus, interfacial tension can be conceived as energy per unit area of interface.
And now, by multiplying interfacial tension by the area of interface, we obtain the
overall energy. Take an O/W emulsion, for example, dispersion of oil droplets in the
aqueous continuous phase increases the surface area of oil (thus the area of oil-water
interface) and causes a rise in energy in the dispersion system as well. Imagine that
we are trying to emulsify 30 mL of oil into 70 mL of water in a container (with
a cross-sectional area of 20 cm²). The smaller the oil droplets become (Fig. 5.9),
the larger is the area of interface. As the droplet diameter drops down to 1 μm, the
interfacial energy rises by 3×10^5-folds (Table 5.2). The more the droplets decrease
in size, the higher the energy would be. In summary, emulsion systems are instable
in terms of energy because emulsification causes increases in interface area and
energy of the systems.

Example 5.2 Four grams of oil and 36 g of water are vigorously agitated with
a homogenizer for 1 min to prepare an O/W emulsion. The oil droplets in the
emulsion have an average diameter, d, of 3.1 μm. The homogenizer consumed 30 W
of electric power for the emulsification operation. Determine the energy efficiency
of the operation. Assume the oil density and the oil-water interfacial tension to be
0.92 g/cm³ and 20 mN/m, respectively.

Solution The 4 g of oil has a volume of $4/0.92 = 4.35$ cm$^3 = 4.35 \times 10^{-6}$ m^3. An oil droplet that measures 3.1 μm in diameter has a volume of $\pi d^3/6 = (3.14)$ $(3.1 \times 10^{-6})^3/6 = 1.56 \times 10^{-17}$ m^3 and a surface area of $\pi d^2 = (3.14)(3.1 \times 10^{-6})^2 = 3.02 \times 10^{-11}$ m^2. The emulsion contains $4.35 \times 10^{-6}/1.56 \times 10^{-17} = 2.79 \times 10^{11}$ oil droplets, contributing to a total oil-water interface area of (3.02×10^{-11}) $(2.79 \times 10^{11}) = 8.42$ m^2. The increase in oil-water interface area is approximated to 8.42 m^2 as the interface area of 4 g of oil is negligible compared to the total interface area after emulsification. We know that the interfacial tension is 20 mN/m $= 0.02$ N/m $= 0.02$ J/m^2, making the increase in interfacial energy (0.02) $(8.42) = 0.168$ J. Meanwhile, the units of electric power are W $=$ J/s, and thus the energy consumption by the homogenizer during the emulsification operation is $(30)(60) = 1800$ J. Merely $0.168/1800 = 9.3 \times 10^{-5} \approx 0.01$ % of the electrical energy consumed is used to increase the interface area. We can see from here that emulsification is a process of low energy efficiency where most of the energy is consumed in the form of heat causing a temperature increase in the process liquid. Approximating the specific heat capacity of the emulsion to that of water $(4180$ J/(kg·K)$)$ and assuming that all the energy spent is consumed in the form of heat, the emulsion will experience a temperature increase of $1800/[(4180)$ $(0.04)] = 10.7$ K $= 10.7$ °C (where K is the unit of temperature difference). ◢

5.3 Rheology

5.3.1 Newton's Law of Viscosity

When we stir a mayonnaise in a container with a spoon, the mayonnaise in the vicinity of the spoon moves along even if there is no direct contact with the moving spoon. A force is needed to move a still object. The mayonnaise that is not in direct contact with the spoon also moves along due to transmission of force from the contacting mayonnaise. The same phenomenon can also be observed in water, but comparatively greater forces are required to stir an apparently "thicker" mayonnaise. The nature of fluids by which a force acts from a fast- to slow-moving layer of the fluids is regarded as *viscosity*.

A layer of water is sandwiched between two horizontal plates, one stationary and one movable in horizontal direction, that are separated at the distance of L [m] (Fig. 5.10). If the upper movable plate is slid by a force, F [N], to move horizontally at velocity, u_0 [m/s], the layer of water adjacent to the moving plate will move the fastest, and each layer of water further away from the moving plate will move slower than the layer above it until the final layer adjacent to the stationary plate stays still. In other words, a velocity distribution occurs as depicted in Fig. 5.10, where the force that transmits from top to bottom per unit area (known as *shear stress*), τ, is proportional to the velocity gradient. Shear stress, τ, if we disregard the directions of the velocity and force, can be expressed by

Fig. 5.10 Simple shear flow
of water sandwiched between
two flat, horizontal plates

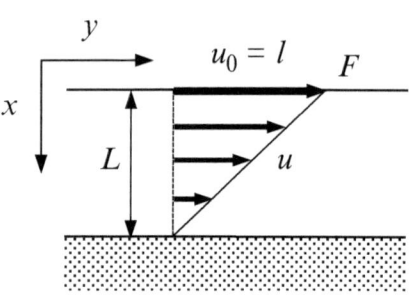

$$\tau = \mu \frac{u_0}{L} \tag{5.2}$$

where the proportionality constant, μ, is the *viscosity* (also known as coefficient of viscosity). Generally, the direction aspect of velocity and force is taken into account. Consider the distance in the direction from the movable plate to the stationary plate as positive; the velocity gradient in Fig. 5.10 will take a negative value. In addition, the force is positive because it is transmitted in the positive direction opposite to that of the velocity gradient. After factoring in the direction aspect of velocity and acting force, the relationship described by Eq. (5.2) can be generalized and expressed as follows:

$$\tau = -\mu \frac{du}{dx} \tag{5.3}$$

where x is the distance and du/dx $[(m/s)/m = s^{-1}]$ is the velocity gradient. Equation (5.3) is called *Newton's law of viscosity*, and fluids of which the coefficients of viscosity are constant irrespective of the magnitude of velocity gradient are called *Newtonian fluids*. Water and glycerol (glycerin) are Newtonian fluids. On the other hand, the coefficients of viscosity of many liquid foods including mayonnaise, as explained later in the chapter, depend on the magnitude of velocity gradient. These fluids are known as *non-Newtonian fluids* which are further classified into different types according to their profiles of dependency (discussed hereinafter).

In Fig. 5.10, let the distance traveled by water in one second (1 s) be l, and then $u_0 = l$ [m/s]. So Eq. (5.2) can be expressed as

$$\tau = -\mu \frac{l}{L} \tag{5.4}$$

where l/L denotes the rightward deformation (strain) of distance, L, per unit time, which is regarded as deformation rate or strain rate. Likewise for Eq. (5.3), let the rightward distance be y and u can be expressed by

$$u = dy/dt \tag{5.5}$$

where t is time. Substituting Eq. (5.5) in Eq. (5.3) gives

$$\tau = -\mu\frac{du}{dx} = -\mu\frac{dy/dt}{dx} = -\mu\frac{dy/dx}{dt} \tag{5.6}$$

where dy/dx represents the rightward minute deformation (strain) with respect to vertically downward distance, dy. In the x-y coordinate system in Fig. 5.10, however, the strain in the y-direction decreases along the positive x-direction, and therefore $dy/dx = -dy$. Expressing $(dy/dx)/dt$—the time derivative of dy/dx—as *deformation rate* (strain rate), $\dot{\gamma}$, Eq. (5.6) can be given by

$$\tau = -\mu\frac{dy/dx}{dt} = \mu\frac{dy}{dt} = \mu\dot{\gamma} \tag{5.7}$$

As been described above, when the velocity of a moving fluid shows a distribution, a force is transmitted from fast- to slow-moving regions. However, to be exact, momentum instead of force, which is given by the product of velocity and mass, is transmitted. As the masses of mayonnaise and water are constant regardless of location, the regions of faster movement contain higher momentum and vice versa; thus momentum transmits from fast- to slow-moving regions. The flow of physical quantities per unit area per unit time is referred to as *flux*, and when the physical quantity refers to momentum, it is called *momentum flux*. The definition of momentum flux is given by

$$\text{Momentum flux} = \frac{(\text{momentum})}{(\text{area})\ (\text{time})} = \frac{(\text{mass})\ (\text{velocity})}{(\text{area})\ (\text{time})} \tag{5.8}$$

Momentum flux is thus presented in the units of

$$\frac{\text{kg} \cdot \text{m/s}}{\text{m}^2 \cdot \text{s}} = \frac{\text{kg} \cdot \text{m/s}^2}{\text{m}^2} = \frac{\text{N}}{\text{m}^2}$$

where m/s^2 are the units of acceleration. The numerator represents the product of mass and acceleration, which is equivalent to force, while the denominator represents the area. Therefore, momentum flux represents stress which is force per unit area. Because the transmission of force occurs in the direction perpendicular to the flow direction of the fluid, seemingly cutting across the fluid flow, it is called *shear stress* in the fields of food science and polymer science.

5.3.2 Measurement of Viscosity

Viscosity is one of the important physical properties of viscous liquid foods, which can be determined with various measurement devices. Figure 5.11 depicts a

Fig. 5.11 A rotational
viscometer

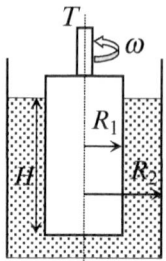

concentric cylinder rotational viscometer (B-type viscometer) which is widely used
for measuring viscosity of liquid foods. The sample to be measured is filled into
the gap between the internal and external cylinders with radii of R_1 [m] and R_2 [m],
respectively. The internal cylinder in contact with the sample measures H [m] in
height. Let the torque acting on the internal cylinder be T [N · m] when it rotates at
the angular velocity of ω [rad/s]. The viscosity of the sample can be determined by

$$\mu = \frac{T}{4\pi\omega H} \left(\frac{1}{R_1{}^2} - \frac{1}{R_2{}^2} \right) \tag{5.9}$$

Torque is a cross product between force and distance (moment). According to this
principle, the force needed to rotate an object varies with the acting point of the
force, and it is generally inversely proportional to the distance of the acting point
from (the center of) the axis. However, the torque is constant irrespective of the
acting point of the force.

5.3.3   Flow Curve

The plot of the shear stress, τ, measured of a fluid as a function of the deformation
rate (shear rate), $\dot{\gamma}$, yields the *flow curve* of the fluid. Newtonian fluids show linear
relationship between the two variables where the straight lines pass through the
origin (Eq. 5.3 or Eq. 5.7) and the slopes tell us the viscosity. However, as mentioned
earlier, many viscid foods are non-Newtonian fluids that do not display linear
relationships between shear stress and deformation rate. Furthermore, some flow
curves also do not pass through the origin.

   Mayonnaise is a non-Newtonian fluid. If we tip a plate with mayonnaise in it up
just a little, the mayonnaise will not flow because the acting force is small. However,
it will start to flow when we further tilt the plate at greater angles. In the case where
application of certain amount of force is necessary before a fluid starts to flow, the
fluid is regarded as having a yield value, and this amount of force (per unit area)
is called *yield stress*, τ_f. Taking yield stress into consideration, the flow curves of
non-Newtonian fluids can be expressed by

Table 5.3 Flow curve of a mayonnaise sample

Shear rate [s^{-1}]	0.24	0.85	1.62	2.87	3.81
Shear stress [Pa]	115	139	160	187	205

Fig. 5.12 The flow curve of a mayonnaise sample

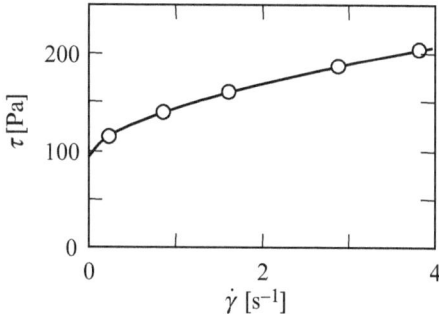

$$\tau = \tau_f + \eta \left(\frac{d\gamma}{dt}\right)^n = \tau_f + \eta \dot{\gamma}^n \qquad (5.10)$$

where η denotes the apparent viscosity and n is the *flow behavior index*. The equation expresses Newtonian flow curves when $n = 1$. Do take note that not all non-Newtonian fluids have a yield stress.

Example 5.3 Shear stress, τ, was measured of a mayonnaise sample at various deformation rates, $\dot{\gamma}$, using a concentric cylinder rotational viscometer as tabulated in Table 5.3. Plot the flow curve and determine the yield stress, the apparent viscosity, and the flow behavior index.

Solution Plotting the shear stress as a function of deformation rate on a normal graph paper yields the flow curve as shown in Fig. 5.12. Next, we will determine the yield stress, τ_f, the apparent viscosity, η, and the flow behavior index, n, from the results. There are a total of three parameters to be determined, but by Eq. (5.10), even appropriately transformed and linearized, we can only obtain two out of the three parameters. Therefore, we have to find out one of the parameters one way or the other. For example, if we manage to determine the yield stress, τ_f, we can transform Eq. (5.10) into

$$\log (\tau - \tau_f) = \log \eta + n \log \dot{\gamma} \qquad (5.11)$$

Therefore, the logarithmic plot of $\tau - \tau_f$ versus $\dot{\gamma}$ yields a straight line with the apparent viscosity, η, determined as the intercept and the flow behavior index, n, as the slope. Knowing that the yield stress, τ_f, is given by the intercept between the flow curve and the y-axis, we can extrapolate the smooth flow curve that connects all the data points in Fig. 5.12 using a French curve (see Fig. 2.9) to estimate τ_f from the y-coordinate of the intercept. Besides, the *Casson equation*, which was originally

Fig. 5.13 Estimation of yield
stress by the Casson equation

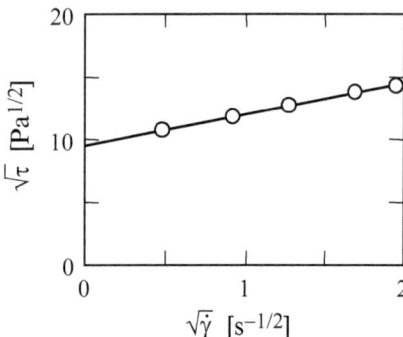

Fig. 5.14 Determination of
apparent viscosity and flow
behavior index

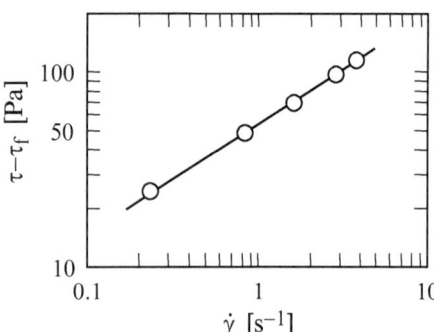

derived to describe oil-based printing inks, has also found use in characterizing the
flow properties of many liquid foods:

$$\sqrt{\tau} = \sqrt{\tau_C} + \sqrt{\mu_C}\sqrt{\dot{\gamma}} \tag{5.12}$$

where τ_C and μ_C denote the Casson yield stress and the Casson viscosity,
respectively. Assuming that τ_C and τ_f are equal, the apparent viscosity, η, and
the flow behavior index, n, can now be obtained from the plot of Eq. (5.11).
The plot of $\sqrt{\tau}$ against $\sqrt{\dot{\gamma}}$ yields a straight line as shown in Fig. 5.13, and τ_C
is determined from the intercept to be $\tau_C = (9.52)^2 = 90.6 = \tau_f$. By substituting
$\tau_f = 90.6$ Pa in Eq. (5.11) and plotting $\tau - \tau_f$ versus $\dot{\gamma}$ on a double logarithmic
graph paper, we obtain Fig. 5.14, and the slope of the straight line gives the flow
behavior index, $n = 0.56$. Further, reading the value of $\tau - \tau_f$ when $\dot{\gamma} = 1$ from the
straight line, the apparent viscosity, η, is estimated to be $\eta = 53.5$ Pa $\cdot$ s$^{0.56}$. The solid
curve in Fig. 5.12 is the computational curve resulting from the substitution of the
determined values in Eq. (5.10). ◢

Generally, fluids with a flow behavior index, $n < 1$, and an upwardly convex flow
curve are called *pseudoplastic fluids*, whereas those with $n > 1$ and a flow curve
concaving downward are regarded as *dilatant fluids*. As mentioned earlier, the flow
curve of a Newtonian fluid is a straight line with $n = 1$. Apart from that, fluids that

Fig. 5.15 Flow curves of
non-Newtonian fluids

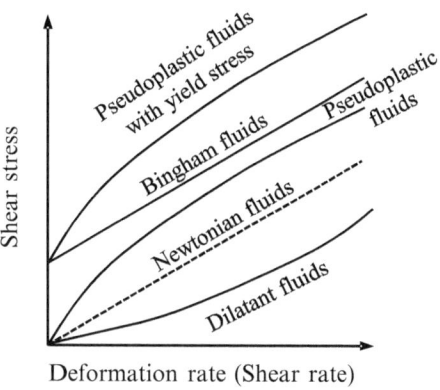

do not flow before the yield stress is reached but behave like Newtonian fluids once
they begin to flow are known as *Bingham fluids*. Meanwhile, fluids like mayonnaise
that exhibit a yield stress above which they behave like pseudoplastic fluids do not
fall under any distinct categories of non-Newtonian fluids, but are just known as
pseudoplastic fluids with yield stress (Fig. 5.15).

5.3.4 Viscoelastic Fluids

(1) *Characteristics of viscoelastic fluids*: *Viscoelastic fluids* are a type of non-
Newtonian fluids that exhibit both Newtonian behavior and elastic characteristic
in their flow characteristics. Being made up of naturally occurring polymer,
a great number of foods are viscoelastic fluids. Whisked egg white, grated
Japanese yam, and agar gel are typical examples of viscoelastic fluids.

Imagine that a step constant stress, τ (force/area), is applied to an object
from time 0 to t as illustrated in Fig. 5.16a. In the case of an elastic solid,
the stress causes a deformation (strain) of γ as depicted in Fig. 5.16b, but the
deformation returns to 0 from γ instantaneously when the stress is removed
after time t. The described relationship between deformation, γ, and stress, τ,
is defined by Hooke's law as

$$\tau = E\gamma \tag{5.13}$$

where E denotes the *elastic modulus* [Pa·s]. When subjected to a similar step
constant stress, the deformation of a viscous fluid increases linearly with time
(the fluid flows corresponding to the stress) as shown in Fig. 5.16c. Even
when the stress is removed after time t, the deformation does not drop back
to 0 but remains constant instead. That is to say that when stress is removed,
elastic solids return to their original shapes and sizes while viscous fluids do
not. Meanwhile, viscoelastic fluids exhibit intermediate combination of both of

Fig. 5.16 Deformation of different materials under a step constant stress

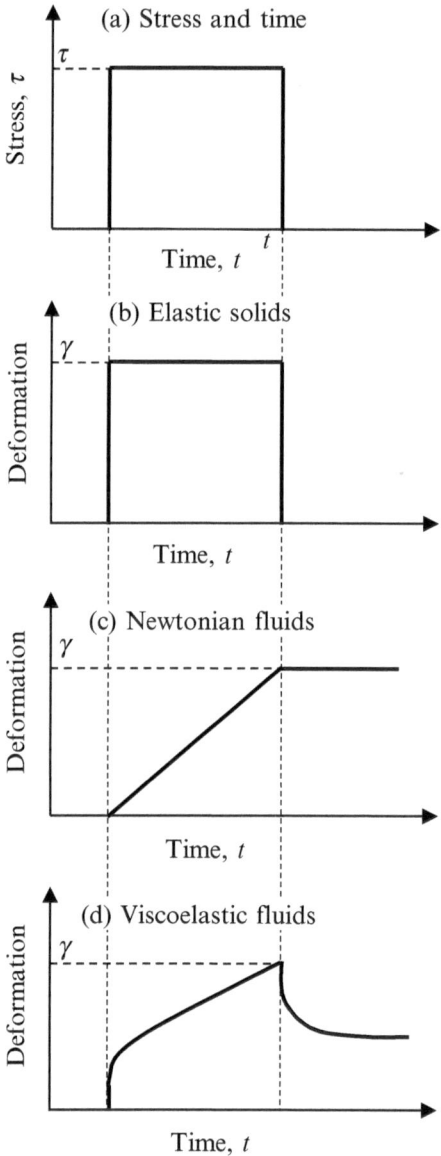

these properties. Figure 5.16d shows the deformation (strain) of a viscoelastic fluid when a step constant stress is applied and the change in deformation when the stress is removed instantaneously. The former change in deformation over time is known as *creep* and the latter is called *creep recovery*. When subjected to a stress, viscoelastic fluids experience elastic deformation (deform instantaneously) as illustrated in Fig. 5.16b, and the deformation eventually increases with time. If the stress is maintained for a sufficiently long period of time, the viscoelastic fluid will start to flow, and the deformation will start increasing linearly with time. When the stress is removed, the deformation decreases immediately to a small extent and continues to drop thereafter at a decreasing rate, slowly approaching a constant residual deformation. Viscoelastic fluids do not return completely to their original shapes and sizes like elastic solids do, but instead behave partially like Newtonian fluids.

(2) *Unique flow behaviors of viscoelastic fluids*: Unique flow conditions which are not seen in any Newtonian fluids are observed when viscoelastic fluids are allowed to flow in a pipe or a container.

   (a) *Recoil flow behavior in a circular pipe*: Imagine a fluid flowing slowly (the type of flow called laminar flow) along a circular pipe, and we manage to dye the fluid front (surface) in some way. In the case of Newtonian fluid, the fluid front will develop from a flat surface gradually into a distribution form similar to a spheroid that rotates about the axis of the pipe. If we now stop the flow in this condition, the fluid front will keep its deformed shape (left diagram in Fig. 5.17a) with the fluid front profile and position of the apex unchanged. This observation is in agreement with the afore-described phenomenon where the deformation, γ, of a Newtonian fluid remains constant even when the stress is removed. Whereas in the case of viscoelastic fluid, although the fluid front profile remains unchanged in shape, the gradual regression of the profile apex is observed. This phenomenon is analogous to that of an extended rubber thread retracting back to its original position when the pulling force is removed, which is attributable to the decrease in the amount of deformation due to the elasticity of the fluid.

   (b) *The Weissenberg effect*: Imagine a vertically suspended cylinder rod immersed and rotating in a fluid contained in a beaker as shown in Fig. 5.17b, and let us observe the flow around the spinning rod. As for Newtonian fluids, the flow rotates concentrically with the rod, and the fluid in the vicinity of the rod moves away from the axis toward the wall of the beaker, causing the fluid surface in the region to sink. On the other hand, in the case of viscoelastic fluids, the fluid in the vicinity of the rod is instead drawn toward the rod and clambers up around it. The phenomenon is called the Weissenberg effect which can be observed in fluid foods like grated Japanese yam.

   (c) *Flow in a semicircular trough*: Imagine that a fluid is flowing down a slanted semicircular trough by a gravitational pull. A Newtonian fluid flows with a flat fluid surface as opposed to a viscoelastic fluid which forms a surface concaving downward. Figure 5.17c illustrates cross-sectional schematics of semicircular troughs with a flowing Newtonian fluid (left) and a viscoelastic fluid (right).

   (d) *Circulation flow in a cylindrical container*: Imagine that a cylindrical container is filled with a fluid and a disk is rotating at the fluid surface. Two types of flows arise in the fluid: (i) the rotational flow just underneath

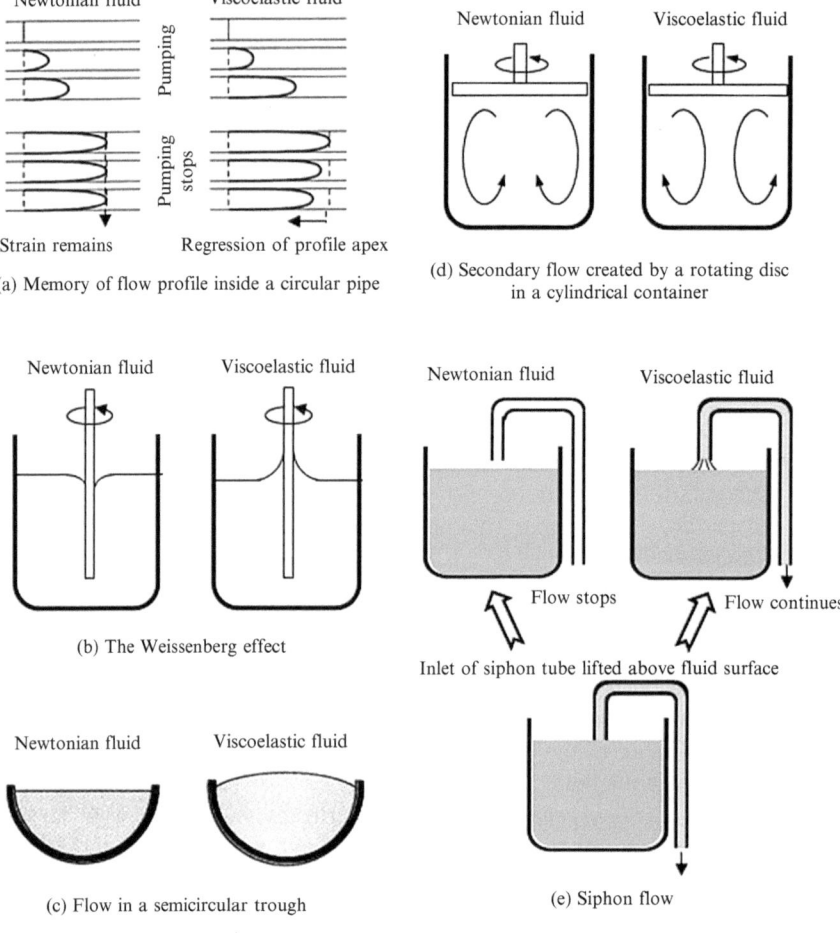

Fig. 5.17 Unique flow behaviors of viscoelastic fluids

the rotating disk and (ii) the secondary axial and radial flow (Fig. 5.17d). In a Newtonian fluid, this secondary flow moves down parallel to the container wall from the rotating disk, then radially inward along the bottom, and eventually upward along the center of the container. On the contrary, the secondary flow in a viscoelastic fluid travels in the exact opposite direction where the fluid travels downward along the rotation axis, then radially outward alongside of the bottom, and finally upward along the stationary container wall.

(e) *Flow in a siphon*: Imagine we are trying to transfer a fluid out of a beaker using a siphon tube. For fluids showing Newtonian behavior, the siphon action stops immediately as the inlet of siphon tube is elevated above fluid

Fig. 5.18 Schematics used for the expression of viscoelastic behaviors of fluids

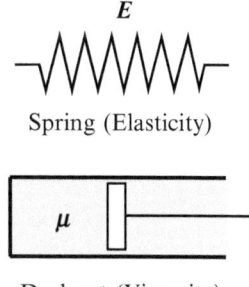

Spring (Elasticity)

Dashpot (Viscosity)

surface. However, when siphoning a viscoelastic fluid, the siphon action will continue even if the siphon tube inlet is to be lifted a few centimeters above fluid surface. This is called the tubeless siphon effect.

(3) *Stress-strain relation of viscoelastic fluids*: Schematics like those illustrated in Fig. 5.18 are used to elucidate the relationship between deformation (strain) and force (stress) in viscoelastic fluids. Elastic deformation (strain) in viscoelastic fluids is depicted by the extension of a spring while viscous flow by the sliding of a piston in a cylinder (called dashpot). The *Maxwell model* and the *Voigt model* are fundamental to the models that describe the behaviors of viscoelastic fluids.

(a) *Maxwell model*: As illustrated in Fig. 5.16, when a fluid is subjected to a stress, τ, the Newton equation (Eq. 5.7) applies for the relationship between stress, τ, and deformation rate, $\dot{\gamma}$, for the case of Newtonian fluids. Meanwhile, Hooke's Law (Eq. 5.13) applies when the fluid is an elastic solid. Differentiating both sides of Eq. (5.13) with respect to time, t, gives

$$\dot{\gamma} = \frac{1}{E}\frac{d\tau}{dt} \qquad (5.14)$$

The Maxwell model, as shown in Fig. 5.19, can be represented by a spring (elasticity) and a dashpot (viscosity) connected in series. Thus, the overall deformation rate, $\dot{\gamma}$, of a viscoelastic fluid can presumably be expressed as the sum of the deformation rates of a Newtonian fluid and an elastic solid, which is given by

$$\dot{\gamma} = \frac{\tau}{\mu} + \frac{1}{E}\frac{d\tau}{dt} \qquad (5.15)$$

When a Maxwell fluid is subjected to the step constant stress depicted in Fig. 5.16a, the fluid instantaneously undergoes an initial deformation, and subsequently the deformation increases over time. Furthermore, part of the deformation decreases at once as the stress is removed, but the residual

Fig. 5.19 Schematic of the
Maxwell model

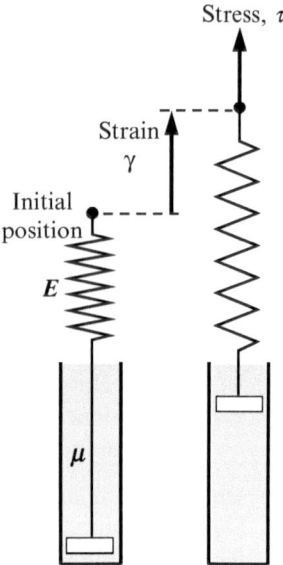

deformation due to viscosity (equivalent to the displacement in the dashpot)
does not recover and is maintained at a constant value.

(b) *Voigt model*: The stress-deformation relationship of an elastic sponge
rubber can be expressed by Hooke's law as in Eq. (5.13). When a sponge
rubber soaked up with a viscous fluid is subjected to deformation, the
resistance force attributed to the viscosity of the liquid will also contribute
to the stress, and the stress-deformation relationship is thus governed by

$$\tau = E\gamma + \mu\frac{d\gamma}{dt} \tag{5.16}$$

The Voigt model can be represented by a spring and a dashpot connected in
parallel (Fig. 5.20). When a step constant force (stress) is exerted on a Voigt
fluid, the spring tends to expand instantly, but the suppressive action by
the dashpot connected in parallel causes the expansion to occur gradually,
preventing an initial abrupt deformation that resembles the one depicted in
Fig. 5.16d. Likewise, the deformation also does not decrease abruptly but
instead recovers slowly.

(c) *Relaxation time*: An object deforms progressively when a force (stress) is
applied on it (Fig. 5.21a). After some time (marking the point in time as 0),
if the stress is maintained at a constant value, the changes in stress thereafter
over time vary with the type of material as shown in Fig. 5.21b. The stress
remains constant (τ_0) in elastic solids (dashed line in Fig. 5.21b), while it
drops instantly to 0 in Newtonian fluids (dot-dashed line in Fig. 5.21b). As
for viscoelastic fluids, the stress undergoes an intermediate decay which

Fig. 5.20 Schematic of the Voigt model

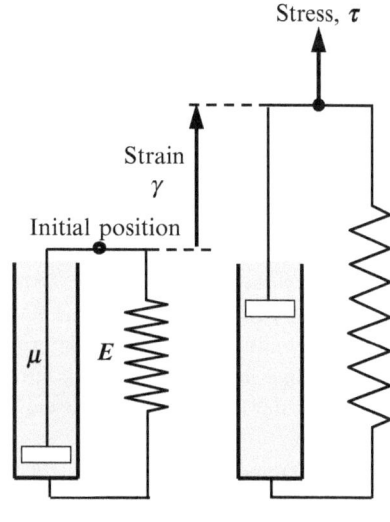

Fig. 5.21 Relaxation time

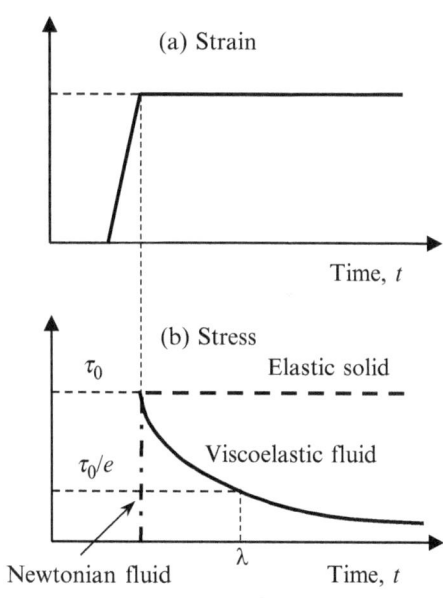

can be represented by the upwardly concaving solid line in the figure. The curve is called a *stress relaxation curve* that depicts the change in stress with time. By the integration of Eq. (5.15) with $\dot{\gamma} = 0$ ($d\gamma/dt = 0$), the curve can be expressed by

$$\tau = \tau_0 \exp\left(-Et/\mu\right) = \tau_0 \exp\left(-t/\lambda\right) \tag{5.17}$$

where $\lambda = \mu/E$ is called *relaxation time* that is defined as the time required for the stress to decay to $1/e$ ($=0.368$). The magnitude of relaxation time of a material varies with the strength of bond between the constituent components of the particular material. Moreover, whether a material behaves like a Newtonian fluid or an elastic solid depends also on the time of observation. The ratio of time of relaxation to time of observation is known as the *Deborah number* (*De*). Materials with $De \ll 1$ are known as fluids, whereas those with $De \gg 1$ are known as elastic solids.

(4) *Dynamic measurement of viscoelastic behavior*: Viscoelastic fluids can basically be characterized by the static method of constructing the creep curves and the dynamic technique of determining the viscoelastic constants from the stress response curves of the fluids subjected to infinitesimal sinusoidal deformations. The latter has been widely used nowadays. We will further explain the principle below.

Imagine that a viscoelastic fluid is subjected to a sinusoidal deformation described by the following equation (for instance, by introducing sinusoidal lateral oscillation through the internal cylinder of the rotational viscometer in Fig. 5.11):

$$\gamma = \gamma_0 \cos \omega t \tag{5.18}$$

where γ_0 is the oscillation amplitude of deformation and ω is the frequency of oscillation. Assuming that viscoelastic fluids are Maxwell fluids and calculating the stress corresponding to the infinitesimal deformation by Eq. (5.15) give $\dot{\gamma} = d\gamma/dt = -\gamma_0\omega \sin \omega t$:

$$\tau = \frac{\mu\omega^2\lambda}{1+\omega^2\lambda^2}\gamma_0 \cos \omega t - \frac{\mu\omega}{1+\omega^2\lambda^2}\gamma_0 \sin \omega t = G'\gamma_0 \cos \omega t - G''\gamma_0 \sin \omega t \tag{5.19}$$

where G' and G'' can be expressed as

$$G' = \frac{\mu\omega^2\lambda}{1+\omega^2\lambda^2} \tag{5.20a}$$

$$G'' = \frac{\mu\omega}{1+\omega^2\lambda^2} \tag{5.20b}$$

Equation (5.19) can be expressed according to trigonometric theorems as follows:

$$\tau = \gamma_0\sqrt{G'^2 + G''^2} \cos (\omega t + \delta) \tag{5.21}$$

Fig. 5.22 Graphical
representation of the principle
of dynamic technique for
characterization of
viscoelastic behaviors

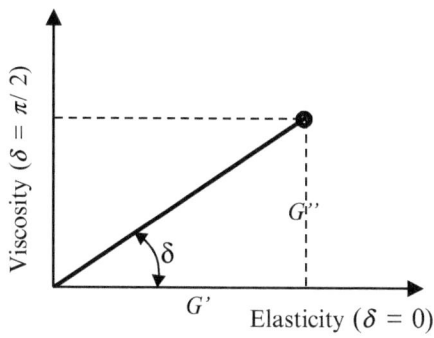

Table 5.4 Concentration, C, and surface tension, γ

Concentration, C [mol/L]	1.0×10^{-6}	2.5×10^{-6}	5.0×10^{-6}	1.0×10^{-5}
Surface tension, γ [mN/m]	72	72	66	56
Concentration, C [mol/L]	3.0×10^{-5}	6.0×10^{-5}	1.0×10^{-4}	2.0×10^{-4}
Surface tension, γ [mN/m]	43	33	30	31

where $\tan \delta = G''/G' = 1/(\omega\lambda)$ and δ is the phase angle called the *loss angle* that expresses the deformation lag in response to stress. $\delta = 0$ is indicative of elastic solids, $\delta = \pi/2$ of Newtonian fluids, and $0 < \delta < \pi/2$ of regular viscoelastic fluids. The numerator, $\mu\lambda$, of G' is equivalent to the elastic modulus, E, which is a parameter that describes elasticity characteristic. The shear modulus, G, may be described using a complex variable as

$$G = G' + iG'' \tag{5.22}$$

where i is the imaginary unit. Plotting G' of $\delta = 0$ on the x-axis against G'' of $\delta = \pi/2$ on the y-axis in a two-dimensional x-y coordinate system, G is presented as a point at the coordinate of (G', G''), at the angle of δ from the x-axis relative to the origin (Fig. 5.22).

Exercise

5.1 Name some example of O/W and W/O emulsions around us.
5.2 Look up the methods for determining the viscosity of fluids.
5.3 The surface tension (at 25 °C) of the aqueous solutions of a surfactant prepared at various concentrations is shown in Table 5.4. What are the critical micelle concentration, the surface excess, and the area occupied by a single surfactant molecule at the interface?

Table 5.5 Flow curve of a tomato ketchup sample

Shear rate [s^{-1}]	3	10	30	100	300
Shear stress [Pa]	6.26	9.34	17.4	32.4	65.3

5.4 Table 5.5 summarizes the shear stress measured of a tomato ketchup sample at prescribed deformation rates (shear rates). Determine the parameters that describe the flow property of the sample.

5.5 Derive Eq. (5.17) that expresses the relaxation curves of viscoelastic fluids.

5.6 A Maxwell fluid was subjected to sinusoidal strains of $\gamma_0 \cos\omega\, t$ using a rheometer to characterize its flow property. At $\omega = 2$ and an angular velocity of 5 s^{-1}, the loss angle, δ, was determined to be 0.375 and 0.158, respectively.
(a) Prove that this set of data is valid and determine the relaxation time, λ.
(b) The oscillation amplitude of τ/γ_0 in Eq. (5.21) was 2.32 kg/(m$^2 \cdot$s) during the measurement at $\omega = 2$ s^{-1}. Determine the viscosity, μ, of the fluid.

Chapter 6
Milk and Vegetables

Abstract The majority of foods are perishables that deteriorate in terms of quality short after being left unattended. We use sterilization of milk as an example to quantitatively elaborate heat sterilization as the most widely used operation for improving storage stability of food. We then discuss the basics of heat transfer and the design of heat exchanger which is a device used for heat sterilization. Refrigeration and freezing are other common practices for extending the shelf life of food. This chapter therefore also documents these operations as well as thawing of frozen food. In view of the close influences water activity has on food storage, the chapter finally closes with discussions on water activity and intermediate-moisture foods using jam as an example.

Keywords Deterioration • Preservation • Sterilization • Pasteurization • High-temperature short-time sterilization • Heat transfer • Heat exchanger • Chilling • Freezing • Thawing • Water activity • Intermediate-moisture food

6.1 Deterioration and Preservation of Food

6.1.1 Quality Deterioration of Food

Pretty much all foods are perishables which deteriorate in terms of quality within short periods of time except for a few others such as table salt. Degraded foods or food ingredients are not only inedible; sometimes they may even pose harm to human body. Quality deterioration in food, depending on the cause, is broadly divided into physical, chemical, biochemical, and biological degradations.

Physical degradation occurs, for example, when a fruit is subjected to an impact and microorganisms invade the fruit through the wounds causing it to rot. In addition, denaturation of proteins due to alteration of the three-dimensional structures caused by heating or freezing and hardening and retrogradation of heat-gelatinized starches at low temperatures are also categorized as physical degradation.

Examples of chemical degradation include denaturation of proteins caused by changes in pH and salt concentration, generative reactions of noxious or repugnant substances such as aldehydes from oxidation of unsaturated fatty acids in lipids, formation of brown substances (called the Maillard reaction or browning reaction)

© Springer Science+Business Media Singapore 2016
T.L. Neoh et al., *Introduction to Food Manufacturing Engineering*,
DOI 10.1007/978-981-10-0442-1_6

by aminocarbonyl reactions that occur at increased temperatures between reducing sugars and amino compounds (such as amino acids and proteins), etc.

Enzymatic browning of grated apple; low-temperature injury or chilling injury that occurs when fruits of tropical and subtropical origin are stored at low temperatures resulting in browning and brown spots; rigor mortis, a state in which the muscles of slaughtered animals become inextensible and relatively rigid and then followed by autolysis or "self-digestion" wherein proteins decompose causing the muscle cells to soften; etc., are examples of biochemical degradation.

Putrefaction and spoilage of foods resulted from decomposition and degeneration of carbohydrates and proteins contained in the foods by microorganisms are an example of biological degradation. The risks for food poisoning increase when one ingests foods contaminated with microorganisms, especially those which are pathogenic. Besides microorganisms, feeding damage by pest insects is also classified as a type biological degradation.

6.1.2 Preservation of Food

Storage stability of food can be improved by eliminating the factors that cause quality deterioration in the food, for example, by killing the microorganisms (sterilization) that contaminate the food and suppressing their proliferation. Sterilization methods can broadly be classified according to application of heat into thermal sterilization and nonthermal sterilization. The method of moist heat sterilization in thermal sterilization, which utilizes hot water and steam to heat and thus kill microorganisms in foods in the presence of moisture, can be subdivided into pasteurization (≤ 100 °C), high-temperature sterilization (≥ 100 °C), and ultrahigh-temperature sterilization (130–150 °C). Besides, the method to heat foods with little moisture at temperatures between 160 and 180 °C for 30 min or longer is called dry heat sterilization. There are also several other heat sterilization methods such as microwave sterilization by which a food product is heated from the inside by microwave radiation, electrical sterilization by which a food product heats up due to its electrical resistance when an electric current is conducted across it, and infrared or far-infrared sterilization by which a food product is irradiated with infrared or far-infrared radiation to heat up the surface.

On the other hand, for nonthermal sterilization, there are chemical sterilization that utilizes chemicals such as hydrogen peroxide and hypochlorous acid and electromagnetic sterilization that kills microorganisms using radiation and ultraviolet rays with high bactericidal activities (250–260 nm). Other than sterilization, storage of foods at low temperatures to suppress microbial growth is also used for extending storage life. As for low-temperature storage, food products can either be refrigerated above the freezing points from −2 to 20 °C or be frozen below the freezing points at the standard freezing temperature of −20 °C.

We can either reduce the water content or suppress the action of water in order to minimize quality deterioration in high-moisture foods. The operation to reduce the amount of water contained in a food to improve its storage stability is known as drying. There are two basic types of drying methods: ① drying by heating by which water in the liquid phase contained in a food is evaporated and eliminated and ② freeze-drying by which the food to be dried along with the water contained within are frozen and the frozen water is sublimated and removed under reduced pressure (refer to Sect.4.4 "Spray Drying and Freeze Drying").

When a moisture-containing food material is pickled with table salt or in seasonings with high salt content (salt curing), the salt will penetrate inside of the food material. The osmotic pressure thus increases leading to dehydration of microorganisms inside the food and reduction in amount of water available for microbial growth. Generally, microbial growth is inhibited between 5 % (w/w) and 10 % (w/w) salt concentration, but the growth of most food spoilage and pathogenic microorganisms is inhibited only at 10 % (w/w) salt concentration and above. Likewise, the method of pickling foods such as fruits with sugar (sucrose) to increase storage stability by decreasing water availability for microbial growth is known as sugar curing. Since sucrose has a molecular weight greater than NaCl and does not dissociate into ions, a sucrose concentration of 60 % or above is generally necessary for enhancing the preservative effect at ambient temperature. The knowledge about water activity is required for understanding the principles of salt curing and sugar curing.

Smoking not only adds appealing flavors to food and promotes development of color, thus improving color and appearance of meat products, but it also improves the storage stability of the smoked food products because many of the components of smoking agent have some antimicrobial and antioxidative properties. In addition, smoking also reduces the moisture content of food.

Respiration and transpiration continue even in harvested vegetables and fruits. Their physiological activities can be suppressed if being stored at low temperatures. Storage life of produce can also be extended by controlled atmosphere storage wherein oxygen concentration in the storage atmosphere is decreased and carbon dioxide concentration increased to reduce respiratory volume. The presence of organic acids in food products will lower the pH value, rendering microbial growth unfeasible. Pickling food in vinegar is an example means for improving the shelf life of food by pH regulation. Besides, other food additives like preservatives, mildewcides, disinfectant, and antioxidants are also added to food products for extending their shelf life.

6.2 Sterilization of Milk

6.2.1 Microbial Death Kinetics

The sterilization conditions of milk such as 130 °C, 2 s, can sometimes be found printed on milk cartons. Let us consider the reasons for the adoption of this high-temperature short-time sterilization condition. Furthermore, we will also discuss the equipment (heat exchanger) for attaining such sterilization conditions.

Fig. 6.1 Survival curve for microbial cells

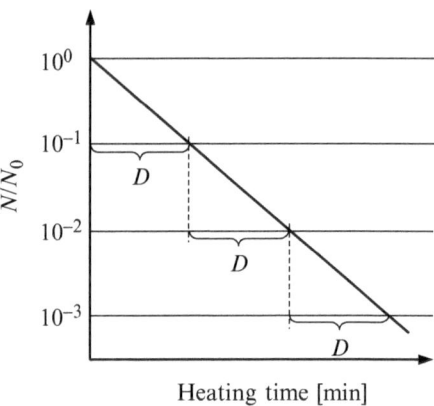

Heating time [min]

As described earlier, sterilization methods may be divided into two main groups: *heat sterilization* by which elimination of microorganisms is achieved by application of heat and *cold sterilization* (aka nonthermal sterilization) which involves the use of chemicals, radiations, etc. The former represents the majority of sterilization operations in food processing, namely, pasteurization (basically 100 °C and below) which targets at vegetative cells and high-temperature sterilization (asepticizing) which targets at spores.

Being subjected to heat treatment at a constant temperature, the number of viable microorganism cells (viable cell count), N, in a food material decreases rapidly over time. Letting the initial viable cell count of microorganisms in liquid food such as milk be N_0, the plot of survival rate, N/N_0, as a function of heating time, t, on a semilogarithmic graph paper often reveals a straight line as shown in Fig. 6.1, which is known as the *survival curve*. When the above-described relationship is obtained, the survival rate can be expressed by either Eq. (6.1) or Eq. (6.2):

$$\frac{N}{N_0} = e^{-k_d t} = \exp\left(-k_d t\right) \tag{6.1}$$

$$\log \frac{N}{N_0} = -\frac{k_d}{2.30} t \tag{6.2}$$

where k_d [s^{-1}] is the *death rate constant*. These equations can be derived based on the consideration of proportionality between the death rate and viable cell count of microorganisms $(dN/dt = -k_d N)$, wherein the microbial death kinetics can be approximated to a first-order reaction. In the food industry, microbial death rate is usually expressed as *D-value* conventionally in the unit of minute, which represents the length of time required for reduction of viable cell count to 1/10 of the initial value. The relationship between k_d and D can be described by

$$D = 2.30/k_d \tag{6.3}$$

Table 6.1 Thermal death process of a microorganism at 95 °C

Heating time [min]	5	15	25	40	60
Survival rate N/N_0	2.82×10^{-1}	1.63×10^{-2}	1.52×10^{-3}	2.22×10^{-5}	1.41×10^{-7}

Fig. 6.2 Survival curve for microbial spores

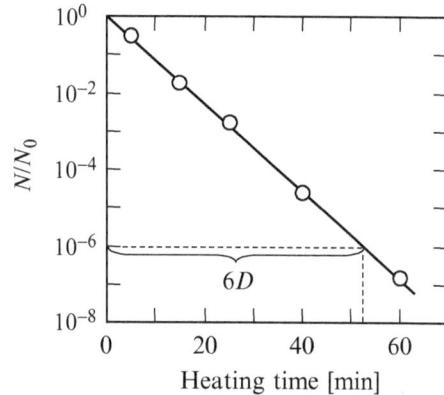

Example 6.1 Table 6.1 summarizes the survival rate at the corresponding heating time for the suspension of spores from a microorganism, while the suspension was being heated at 95 °C. Determine the D-value and k_d.

Solution We will obtain Fig. 6.2 by plotting the data in Table 6.1 on a semilogarithmic graph paper. The value on the horizontal axis (heating time) for N/N_0 to decrease by an order of magnitude is equal to the D-value which can be determined by dividing the value on the horizontal axis for N/N_0 to decrease by n orders of magnitude by n. Tracing along the straight line that passes through the y-intercept of 1 (= 10^0) and reading from the graph, it took 52.2 min for N/N_0 to drop down to a value 6 orders of magnitude less than 10^0 (N/N_0 reduces from 10^0 to 10^{-6}), which is equivalent to $6D$. Hence, $D = 52.2/6 = 8.70$ min. Then from Eq. (6.3), $k_d = 2.30/D = 2.30/8.70 = 0.264$ min^{-1}. Note that k_d can also be determined as described below. The difference in terms of value of the vertical axis when N/N_0 decreases from 10^0 to 10^{-6} is $\log 10^{-6} - \log 10^0 = -6 - 0 = -6$. Meanwhile the corresponding time lapse has been determined earlier to be 52.2 min, so the slope of the straight line in Fig. 6.2 would be $-k_d/2.30 = -6/52.2 = -0.115$ min^{-1}. Therefore, $k_d = (-0.115)(-2.30) = 0.264$ min^{-1}. ◀

6.2.2 Temperature Dependence of Microbial Death Rate

The plot of D-value as a function of temperature, T [°C], on a semilogarithmic graph paper often reveals a linear relationship (Fig. 6.3) known as the *thermal resistance curve* (or thermal destruction curve). The temperature difference corresponding with

Fig. 6.3 Thermal resistance
curve

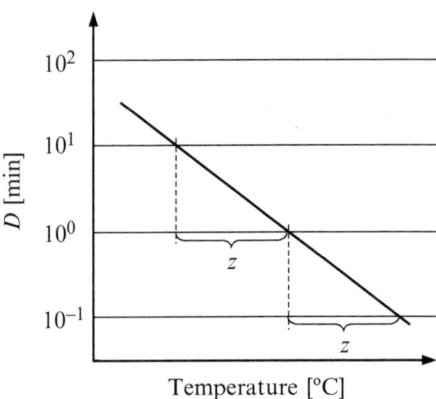

the reduction by one order of magnitude (1/10) in the D-value is called the z-*value* [°C]. The z-value for moist heat sterilization of vegetative cells ranges between 4 and 7 °C, while that of spores varies from 8 to 13 °C.

As been discussed previously, the death rate constant is the rate constant when the microbial death kinetics is assumed to be first order. The temperature dependence of the rate constants of chemical reactions is usually expressed by the *Arrhenius equation*. Hence by applying the Arrhenius equation, the temperature dependence of death rate constant, k_d, can be given by

$$k_d = A_d e^{-E_d/(RT)} = A_d \exp\left(-\frac{E_d}{RT}\right) \tag{6.4}$$

where E_d [J/mol] represents the *activation energy* of the microbial death kinetic process, A_d is the *frequency factor* (or *pre-exponential factor*), T is the absolute temperature in kelvins (K) instead of temperature in degrees Celsius (°C), and R (=8.31 J/mol $\cdot$ K) is the gas constant. Taking the logarithm of both sides of Eq. (6.4) gives

$$\log k_d = -\frac{E_d}{RT}\log e + \log A_d = -\frac{E_d}{2.30R}\frac{1}{T} + \log A_d \tag{6.5}$$

Hence, the semilogarithmic plot of the death rate constant, k_d, obtained by measurements at various (absolute) temperatures against the reciprocal of absolute temperature $(1/T)$ yields a straight line from the slope of which the activation energy, E_d, can be determined. Further, the value of frequency factor, A_d, can be determined by reading the coordinate of an arbitrary point on the straight line and substituting the values into Eq. (6.5).

Example 6.2 The same spore suspension as that mentioned in Example 6.1 was subjected to heat treatment at 90, 92.5, and 97.5 °C. The survival rate was measured and tabulated in Table 6.2. Determine the death rate constant, k_d, at

Table 6.2 Thermal death processes of a microorganism at various temperatures (survival rate)

Heating time [min]	5	10	15	20	25	30
90.0 °C	–	–	–	1.64×10^{-1}	–	–
92.5 °C	–	1.81×10^{-1}	–	–	–	5.82×10^{-3}
97.5 °C	5.12×10^{-2}	2.28×10^{-3}	1.34×10^{-4}	5.26×10^{-6}	2.78×10^{-7}	

Heating time [min]	50	70	80	90	110	140
90.0 °C	1.03×10^{-2}	–	7.32×10^{-4}	–	4.64×10^{-5}	3.28×10^{-6}
92.5 °C	2.11×10^{-4}	6.51×10^{-6}	–	2.46×10^{-7}	–	–

each temperature, and then determine the corresponding activation energy, E_d, and frequency factor, A_d, of the death kinetic process of the particular microorganism by Eq. (6.4) based on the obtained k_d values.

Solution Similar to Example 6.1, Fig. 6.4 can be obtained by plotting the survival rate against heating time for the different temperatures on a semilogarithmic graph paper. The respective death rate constants at 90, 92.5, and 97.5 °C are determined from the slopes of the straight lines to be 0.0901, 0.169, and 0.606 min^{-1}. In addition, we know from Example 6.1 that the death rate constant at 95 °C is 0.264 min^{-1}. The semilogarithmic plot of these rate constant values versus the reciprocal of heating temperature (absolute temperature) is shown in Fig. 6.5, from which the activation energy, E_d, can be determined from the slope of the straight line. However, since the figure uses two different scales for its axes, the slope of the straight line is not simply given by the length ratio of the line segment of AC to that of BC. Instead, the slope can be calculated from the coordinates of any two arbitrary points sitting on that straight line. The line is extended to point A as shown by the dashed line for ease of reading from the y-axis, and the coordinates of points A and B read $(2.681 \times 10^{-3}, 1.0)$ and $(2.749 \times 10^{-3}, 0.1)$, respectively. The change in the y-direction is $\log 0.1 - \log 1.0 = -1 - 0 = -1$, while the corresponding change in the x-direction equals $(2.749 - 2.681) \times 10^{-3} = 6.8 \times 10^{-5}$. Therefore, the slope $-E_d/(2.30R)$ is equal to $-1/(6.8 \times 10^{-5}) = 1.47 \times 10^4$ and $E_d = (1.47 \times 10^4)(2.30)(8.31) = 2.81 \times 10^5$ J/mol $= 281$ kJ/mol.

Then, by substituting E_d and the coordinate of point A into Eq. (6.5), we obtain

$$\log 1 = -\frac{2.81 \times 10^5}{(2.30)(8.31)} 2.681 \times 10^{-3} + \log A_d$$

$$\log A_d = 0 + 39.4 = 39.4$$

$$A_d = 10^{39.4} = 2.51 \times 10^{39} \, \text{min}^{-1} \blacktriangle$$

Fig. 6.4 Survival curves of microbial spores at 90.0 °C (○), 92.5 °C (△), and 97.5 °C (□)

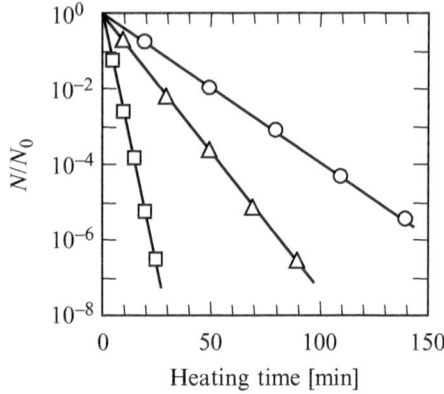

Fig. 6.5 Arrhenius plot of estimated death rate constants, k_d

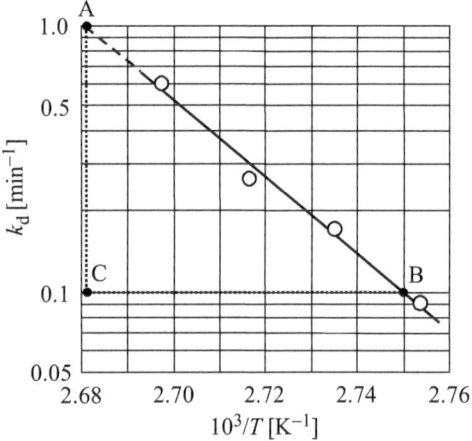

6.2.3   Flash Pasteurization (High Temperature and Short Time)

During the sterilization process of foods, not only microorganisms are eliminated, but the compounds contained in the food materials are also subjected to the influence of heat. Some of the heat-induced reactions therein such as the decomposition of useful compounds are undesirable. Assuming that these are also first-order reactions, let us consider the case in which the temperature dependence of the reaction rate constant can also be expressed by the Arrhenius equation similar to the microbial death kinetic process. Hence, the relationship between the concentration of useful food compound, C, and heating time, t, can be expressed by

$$C/C_0 = e^{-k_b t} = \exp\left(-k_b t\right) \tag{6.6}$$

Table 6.3 Thermal decomposition of vitamin B_1 at various temperatures (residual ratio)

Heating time [min]	60	90	180	270	360	450
90 °C	0.946	0.931	0.843	0.802	0.718	0.679
98 °C	0.886	0.831	0.716	0.608	0.493	0.435
107 °C	0.754	0.649	0.447	0.294	0.186	0.131
115 °C	0.552	0.424	0.168	–	–	–

where C_0 denotes the initial concentration of the useful food compound and k_b [s^{-1}] represents the decomposition rate constant of the compound, the temperature dependence of which can be described by

$$k_b = A_b e^{-E_b/(RT)} = A_b \exp\left(-\frac{E_b}{RT}\right) \tag{6.7}$$

where E_b [J/mol] is the activation energy for the decomposition reaction of the compound and A_b [s^{-1}] is the frequency factor.

Example 6.3 Table 6.3 shows the residual ratio, C/C_0, of vitamin B_1 when the aqueous solutions are heated at different temperatures. Determine the decomposition rate constant, k_b, of vitamin B_1 at each temperature. Also, determine the activation energy, E_b, for the thermal decomposition of vitamin B_1 in the aqueous solution and the frequency factor, A_b.

Solution We obtained Fig. 6.6 by plotting the residual ratio, C/C_0, of each temperature as a function of heating time on a semilogarithmic graph paper. The plots for all temperatures revealed straight lines, indicating that the decomposition kinetics at all temperatures approximate to first order. Hence, the decomposition rate constants, k_b at 90, 98, 107, and 115 °C, are determined from the slopes of the corresponding straight lines by Eq. (6.6) to be 8.76×10^{-4}, 1.89×10^{-3}, 4.57×10^{-3}, and 9.84×10^{-3} min^{-1}, respectively. Then, by plotting the decomposition rate constant against the reciprocal of heating temperature (absolute temperature), we obtained the straight line shown in Fig. 6.7, and the activation energy, E_b, for the thermal decomposition of vitamin B_1 is subsequently determined from the slope by Eq. (6.7) to be 1.14×10^5 J/mol $= 114$ kJ/mol. Plugging the coordinate of an arbitrary point on the straight line and the obtained E_b value into Eq. (6.7), the frequency factor, A_b, is calculated as 2.06×10^{13} min^{-1}. ◢

The activation energy for the death process of the microorganism (spores) determined in Example 6.2 is far greater than that for the decomposition process of the functional food compound obtained in Example 6.3. It is a general trend that the activation energy for the death process of spores is several times larger than that for the thermal degradation (decomposition) of food compounds, indicating that the death process of spores is susceptible to the influence of temperature. The activation energy for the death process of vegetative cells is even significantly greater where the cells can be easily killed with a slight rise in heating temperature.

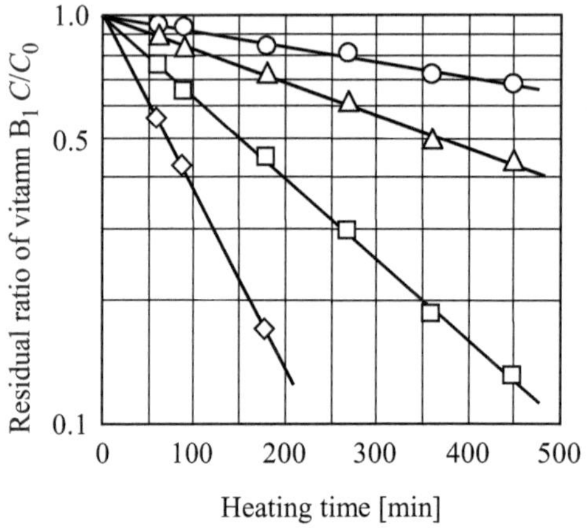

Fig. 6.6 Decomposition of
vitamin B$_1$ at 90 °C (O),
98 °C (△), 107 °C (□), and
115 °C (◇)

.

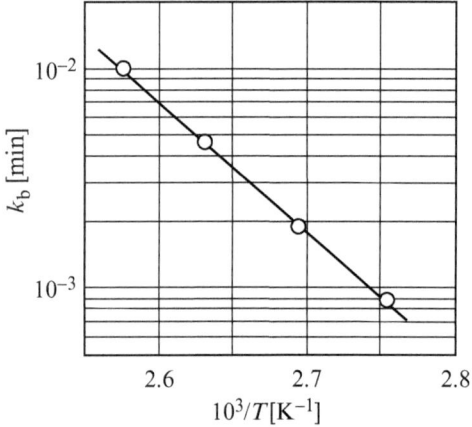

Fig. 6.7 Arrhenius plot of
estimated decomposition rate
constants

Let us consider the effect of sterilization temperature on the degradation of food compounds based on the difference in activation energy between the death process of microorganisms and the degradation process of food compounds. The death rate constants of microorganisms at temperatures T_1 and T_2 ($T_2 > T_1$) vary greatly. On the other hand, although the degradation rate of food compounds does increase with temperature, the difference is not so big because of the low activation energy. The survival rate, N/N_0, of microorganisms and the residual ratio, C/C_0, of food compounds at T_1 and T_2 change over time as depicted by the solid line and dashed line, respectively, in Fig. 6.8. In food processing, the survival rate of microorganisms must be reduced below a certain level. Heating at the higher temperature of T_2 for the length of time of t_2 would be required to achieve the survival rate of $(N/N_0)_2$, and the ratio of the remaining food compounds at this point in time is given by $(C/C_0)_2$. Meanwhile, the same or lower levels of survival rate must be achieved even by sterilization at the lower temperature of T_1, meaning $(N/N_0)_2 = (N/N_0)_1$.

Fig. 6.8 Microbial death (——) and degradation of food compounds (- - - - -) at T_1 and T_2 ($T_2 > T_1$)

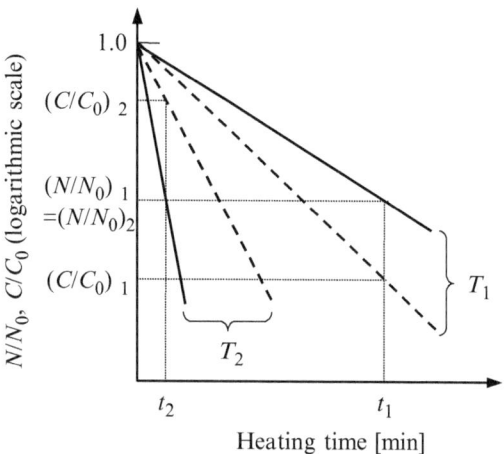

This particular level of survival rate could be achieved in t_1 of time by heating at temperature T_1, and the residual ratio of food compounds at this point is given by $(C/C_0)_1$ which is far lower in terms of value compared to $(C/C_0)_2$. That means to say that far more food compounds will remain when sterilization is performed at higher temperatures to attain a prescribed survival rate (sterilization rate) compared to that carried out at lower temperatures to achieve the same survival rate. For this reason, the high-temperature short-time method has been adopted, for example, in milk processing.

Example 6.4 A vitamin B_1-fortified drink has been contaminated with spores from the same microorganism mentioned in Example 6.2. The drink is subjected to heat sterilization to reduce the survival rate of the spores to one millionth ($X/X_0 = 10^{-8}$) or below of the initial value. What are the respective residual ratios of vitamin B_1 after sterilization at 85 and 121 °C?

Solution The death rate constant at the heating temperature of 121 °C, $k_{d,121} = 2.51 \times 10^{39} \ \exp[-2.81 \times 10^5/(8.31) \ (273 + 121)] = 133 \ \text{min}^{-1}$. The heating time required to attain a survival rate of 10^{-8} is determined from Eq. (6.1) to be $t_{121} = \ln(N/N_0)/(-k_{d,121}) = \ln 10^{-8}/(-133) = 0.138 \ \text{min} = 8.3 \ \text{s}$. The decomposition rate constant of vitamin B_1 at 121 °C, $k_{b,121}$ is equal to $2.06 \times 10^{13} \ \exp[-1.14 \times 10^5/(8.31) \ (273 + 121)] = 0.0156 \ \text{min}^{-1}$, and hence from Eq. (6.6), $C/C_0 = \exp[(-0.0156) \ (0.138)] = 0.998$, indicating that there is almost no decomposition occurring. Meanwhile, at 85 °C, $k_{d,85} = 0.0239 \ \text{min}^{-1}$, and the drink needs to be heated for 771 min to bring the survival rate down to 10^{-8}. Further the decomposition rate constant at 85 °C, $k_{b,85} = 4.70 \times 10^{-4} \ \text{min}^{-1}$ and thus is the residual ratio of vitamin B_1, $(C/C_0)_{85} = \exp[(-4.70 \times 10^{-4}) \ (771)] = 0.696$, indicating a decomposition of approximately 30 %. ◢

6.2.4   Heat Exchanger

Heating and cooling are frequently used operations for manufacturing of food. As discussed in the previous section, heat sterilization of food employs high-temperature fluids (gases and liquids are collectively referred to as fluids) as external heat source, and heat energy is transferred from hot to cold fluids. This phenomenon is known as *heat exchange* and a device therefore is called a *heat exchanger*. Since thermal energy is expensive and thus its efficient consumption is definitely favorable, it is important to develop a deep understanding about the transfer of heat energy.

As shown in Fig. 6.9, there are many different types of heat exchangers, but from the safe point of view, just about all the heat exchangers used for heating and cooling food products are of the type of which the heating media are separated from the process materials by a solid wall (made of metal, plastic, etc.). The most common type of heat exchangers used for heating and cooling of liquid foods are plate heat exchangers (Fig. 6.9a) and tubular heat exchangers (Fig. 6.9b, c).

Plate heat exchangers (Fig. 6.9a) are used in the processing of dairy products and beverages. These exchangers consist of many closely stacked metal plates made commonly of stainless steel, forming parallel flow channels that alternate hot and cold fluids for heat transfer to take place. The surface of the plates is stamped with particular patterns such that turbulent flow is generated and maintained in the fluids to maximize the efficiency of heat transfer. Plate heat exchangers are used for less viscous liquid foods (<5 Pa $\cdot$ s). Further, a particle size of 3 mm or below is preferable when a process fluid contains solids. Prevention of fouling phenomena or sticking and accumulation of components such as milk protein on heat exchange surface is crucial for the application of heat exchanger to milk processing. One of the advantages of this type of heat exchangers is that they can be disassembled easily for cleaning, making them suitable for use for sterilization of dairy products.

Tubular heat exchangers are the most widely used type of heat exchangers. Double-pipe heat exchangers are simple tubular heat exchangers consisting of a smaller tube mounted concentrically to a larger tube (Fig. 6.9b), by which heat exchange takes place as hot and cold fluids or vice versa flow through the annular space and the inner tube, respectively. This type of heat exchangers is employed for sterilization of liquid foods as well because of the simple structures and the ease of cleaning. Shell and tube heat exchangers are the most common type of heat exchangers (Fig. 6.9c). They are used for preheating and heat recovery purposes in the concentration process of liquid foods. Although they have simple structures, these heat exchangers are disadvantaged by the difficulties with cleaning because there exists a lot of dead space along the flow path.

In addition, scraped-surface heat exchangers are another type of heat exchangers in which fouling layers of sugar, protein, etc., are removed periodically from the heat transfer surface, making them suitable also for heating or cooling fluids of high viscosity.

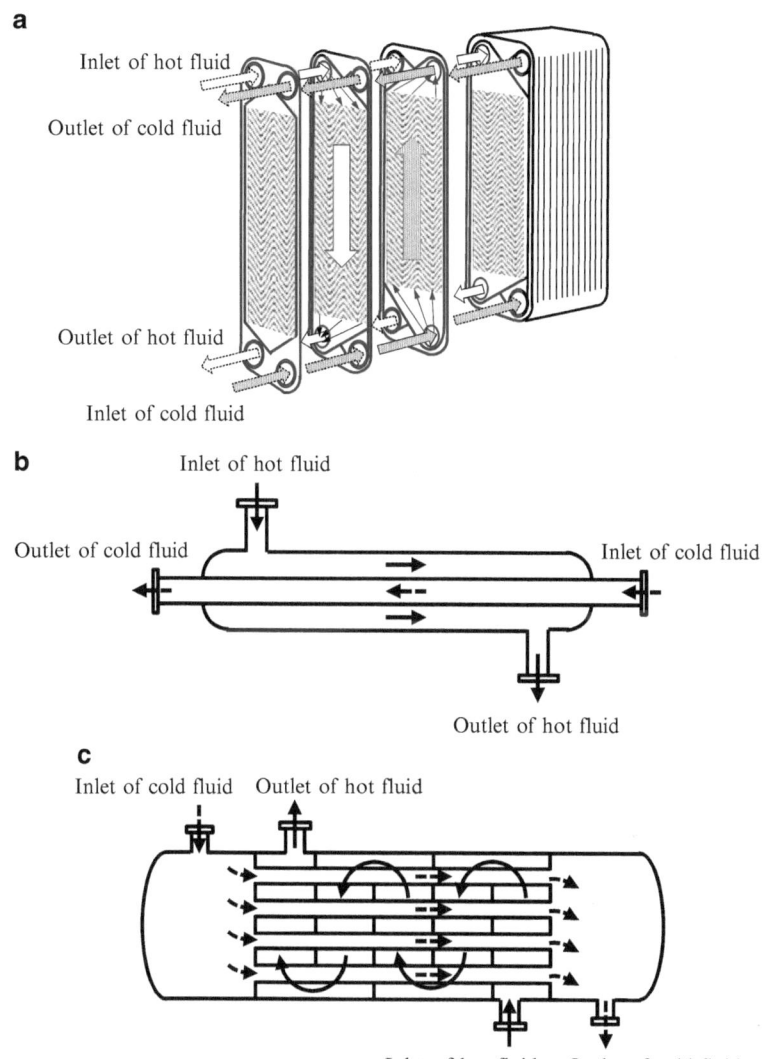

Fig. 6.9 Heat exchangers: (**a**) plate heat exchanger, (**b**) double-pipe heat exchanger, and (**c**) shell and tube heat exchanger

(1) *Transfer of heat:* It is necessary to understand how heat is transferred between two fluids in order to determine the type and size of heat exchanger. Thermal conduction, convection, and radiation are three fundamental modes by which heat energy is transferred. We will discuss thermal conduction and convection here because of their importance in food manufacturing.

Fig. 6.10 Conduction of heat
in solid

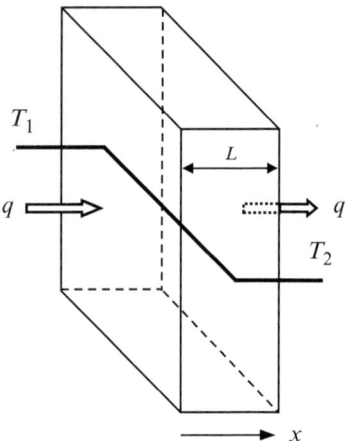

(a) *Heat conduction and thermal resistance*: When a temperature difference
occurs between two points in a solid material, we know even from
experience that heat will flow (transfer) from the hot point to the cold
point. This particular mode of heat transfer is regarded as *conductive heat
transfer* (heat conduction). The principle faces of a plane plate of thickness
L are maintained at temperatures T_1 and T_2 ($< T_1$), respectively, as pictured
in Fig. 6.10. The heat flow, q [W/m^2 ($= $ (J/s)/m^2)], perpendicular to the
principle faces (in the x-direction) per unit area per unit time (*heat flux*) can
be expressed by *Fourier's law* as

$$q = -k\frac{dT}{dx} \tag{6.8}$$

where k [W/(m $\cdot$ K)] is the *thermal conductivity* of the plane plate, of which
higher values indicate easier conduction of heat. Further, dT/dx represents
the slope of temperature distribution (temperature gradient) in the thickness
direction of the plane plate (the x-direction). The negative sign (minus)
on the right-hand side of Eq. (6.8) is derived from the rule of thumb that
heat flows from a higher to a lower temperature (in a direction opposite
to the temperature gradient). In a state where the temperature distribution
does not change over time (steady state), q is a constant value regardless
of x (otherwise, fluctuation of heat energy at a particular point along the
x-direction will occur and a steady temperature distribution will not be
achieved). At this point in time, the temperature gradient inside the plane
plate varies linearly in the x-direction as pictured in Fig. 6.10. Transposing
Eq. (6.8) to $qdx = -kdT$ and integrating within the limits of $T = T_1$ when
$x = 0$ and $T = T_2$ when $x = L$ give

Fig. 6.11 Analogy between (**a**) heat transfer and (**b**) Ohm's law

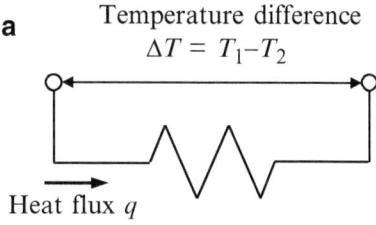

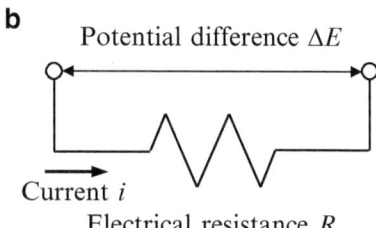

$$\int_0^L q\,dx = -k \int_{T_1}^{T_2} dT$$

$$qL = k\left(T_1 - T_2\right) \qquad (6.9)$$

Transposing Eq. (6.9) gives

$$q = k\frac{T_1 - T_2}{L} = \frac{T_1 - T_2}{L/k} \qquad (6.10)$$

The heat flux, q, in Eq. (6.10) is proportional to the driving force of heat transfer which is the temperature difference, $T_1 - T_2$, and is inversely proportional to L/k. These relationships are analogous to those stated by Ohm's law in Eq. (6.11) (Fig. 6.11a, b): the electric current flowing in a circuit, I, is proportional to the potential difference, ΔE, and is in reverse proportion to the electrical resistance, R.

$$I = \frac{\Delta E}{R} \qquad (6.11)$$

Since L/k of Eq. (6.10) corresponds to the electrical resistance, R, of Eq. (6.11) and represents the ability of a material to resist heat flow, it is thus called the thermal resistance.

Fig. 6.12 Heat conduction
through cylindrical wall

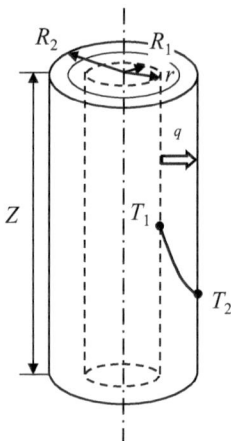

Next, let us consider the heat conduction from the inner wall to the outer
wall of a cylindrical pipe of inner radius R_1, outer radius R_2, and length Z
(Fig. 6.12). Similar to Eq. (6.8), the radial heat flux, q, at r from the center
of the pipe can be expressed by

$$q = -k\frac{dT}{dr} \tag{6.12}$$

Because the area of the surface perpendicular to the heat transfer direction
(heat transfer area), A, in a cylindrical pipe varies with r, the heat flux, q,
is not a constant value as opposed to that for a flat plate. Letting the pipe
length be Z, A is given by $2\pi rZ$; thus the total amount of heat, Q [J/s],
transferred from the inner wall of the pipe to the outer wall through the
cylindrical surface of radius r can be expressed by

$$Q = Aq = 2\pi rZq = -2\pi rZk\frac{dT}{dr} \tag{6.13}$$

In a steady state, Q is constant irrespective of the position within the
cylindrical wall. In the flat plate case, the heat flux, q, is a constant value that
shows no dependence on x; in the cylindrical wall case, the total amount of
heat transferred, $q \cdot A$, is constant independent of r. Transposing Eq. (6.13)
and separating the variables yield

$$\frac{Q}{2\pi rZ}dr = -kdT \tag{6.14}$$

Integrating the left side of Eq. (6.14) with respect to r from R_1 to R_2 and
integrating the right side with respect to T from T_1 to T_2 yield

$$\frac{Q}{2\pi Z}\int_{R_1}^{R_2}\frac{dr}{r} = -k\int_{T_1}^{T_2} dT$$

$$\frac{Q}{2\pi Z}\ln\frac{R_2}{R_1} = k\,(T_1 - T_2) \tag{6.15}$$

Hence, the total amount of heat transferred, Q, can be expressed by

$$Q = \frac{2\pi Z k\,(T_1 - T_2)}{\ln\,(R_2/R_1)} = \frac{T_1 - T_2}{\ln\,(R_2/R_1)\,/\,(2\pi Z k)} \tag{6.16}$$

The right side of Eq. (6.16) is analogous to Eq. (6.10); therefore for heat conduction within cylindrical wall, the thermal resistance can be expressed as $\ln(R_2/R_1)/(2\pi Z k)$.

(b) *Heat convection and heat transfer coefficient*: During a process in which cold milk is heated with high-temperature water vapor (steam), the two fluids are fed through separate flow channels partitioned with solid walls made of stainless steel. Heat transfers across the solid walls from the hot fluid to the cold one as heat exchange takes place. Heat transfer associated with mass motion of a fluid is known as *convective heat transfer* (or shortly as convection). Figure 6.13a depicts the heat transfer from a hot fluid of temperature T_h to a cold fluid of temperature T_c across a solid wall (plate) of thickness L that separates the two fluids. Let the temperature of the plate surface in contact with the hot fluid be T_{i1} ($< T_h$). Assuming that the fluid is flowing at a sufficiently high speed, a layer of fluid in which an abrupt temperature change (a temperature drop in this case) from T_h to T_{i1} is induced forms extremely close to the plate surface. This particular layer is called a *thermal boundary layer* in which fluid flow is not disturbed and thus heat transfer may approximately be perceived as taking place by conduction alone. Letting the thermal boundary layer thickness be δ_h, the heat flux, q, of the layer can be expressed similarly to Eq. (6.10) by

$$q = \frac{T_h - T_{i1}}{\delta_h/k_h} = \frac{k_h}{\delta_h}\,(T_h - T_{i1}) \tag{6.17}$$

where k_h represents the thermal conductivity of the hot fluid. Since the thickness, δ_h, of thermal boundary layer varies intricately with the flow condition of fluid, the pattern on solid wall surface, the properties of fluid (viscosity, density, etc.), and the theoretical determination of δ_h are not possible. So we have no choice but to measure it experimentally. Denoting k_h/δ_h on the right side of Eq. (6.17) as h_h, the heat flux, q, can be expressed by

$$q = h_h\,(T_h - T_{i1}) \tag{6.18}$$

Fig. 6.13 (**a**) Heat transfer between two fluids separated by a solid wall, (**b**) heat resistance, and (**c**) a circuit equivalent to that of (**b**)

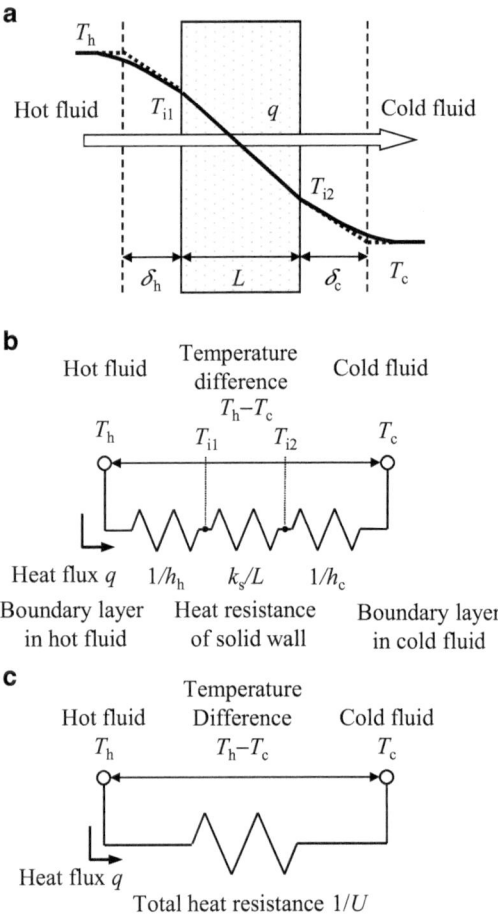

This equation is called *Newton's law of cooling*, and h_h [W/(m^2·K)] is the *heat transfer coefficient* of the thermal boundary layer on the hot fluid side. Comparing with the aforementioned Ohm's law, $1/h_h$ represents the thermal resistance. The heat flux of the thermal boundary layer on the cold fluid side can also be expressed by an equation similar to Eq. (6.18) using h_c to denote the heat transfer coefficient of that thermal boundary layer. Heat transfer coefficients, h_h and h_c, have been determined under a variety of conditions, and empirical formulas have been created for their estimation.

The heat transferred to the solid wall from the hot fluid flows by conduction across the solid wall, then to the thermal boundary layer on the cold fluid side, and finally to the bulk of the cold fluid through convection. The following equation holds because the heat fluxes, q, in the thermal boundary layer on the hot fluid side, the solid wall, and the thermal boundary layer on the cold fluid side are all equal in steady state:

$$q = h_h \left(T_h - T_{i1}\right) = k \left(T_{i1} - T_{i2}\right)/L = h_c \left(T_{i2} - T_c\right) \tag{6.19}$$

where k denotes the thermal conductivity of the solid wall. T_{i1} and T_{i2} of Eq. (6.19) are the interfacial temperatures which are usually not measureable, ending up rendering the calculation of the heat flux, q, not possible. As pictured in Fig. 6.13b, $1/h_h$, L/k, and $1/h_c$ are the respective thermal resistances for the thermal boundary layer on the hot fluid side, the solid wall, and the thermal boundary layer on the cold fluid side, which are regions arranged in series. Expressing the total resistance $(1/h_h + L/k + 1/h_c)$ as $1/U$, Fig. 6.13c can be constructed as a circuit schematic equivalent to Fig. 6.13b. Then, q can be expressed by the following equation with the difference between the measurable temperatures of the hot and cold fluids, $T_h - T_c$ as the driving force:

$$q = U \left(T_h - T_c\right) \tag{6.20}$$

where U [W/(m$^2 \cdot$ K)] denotes the *overall heat transfer coefficient* which is the reciprocal of the total thermal resistance that can be described by the following relationship:

$$\frac{1}{U} = \frac{1}{h_h} + \frac{1}{k/L} + \frac{1}{h_c} \tag{6.21}$$

In the case where heat exchange is performed by heat convection through the tube wall of a tubular heat exchanger (Fig. 6.9b, c), we will have to take into account the curvature of the heat transfer surface similar to the heat conduction through cylindrical walls when considering the heat transfer process. Imagine that a hot fluid and a cold fluid are flowing on the outside and inside of a circular tube, respectively, as depicted in Fig. 6.14. Now, let us consider the heat transfer in this system assuming that thermal boundary layers are present at the interfaces between the fluids and the solid wall. The quantity of heat, Q, that transfers from a hot fluid to a cold fluid per unit time is given by the product of heat flux, q, and heat transfer area, A, and it is constant regardless of the position on the heat transfer surface in steady state as described previously by Eq. (6.13). The respective quantities of heat, Q, that flow across the thermal boundary layer on the hot fluid side (the outer surface of the circular tube), the solid wall, and the thermal boundary layer on the cold fluid side (the inner surface of the circular tube) pictured in Fig. 6.14 can be expressed using Eq. (6.18) and Eq. (6.16) by

$$Q = 2\pi R_2 Z h_h \left(T_h - T_{i1}\right) = 2\pi Z k \left(T_{i1} - T_{i2}\right)/\ln\left(R_2/R_1\right)$$
$$= 2\pi R_1 Z h_c \left(T_{i2} - T_c\right) \tag{6.22}$$

Fig. 6.14 Heat exchange between fluids separated by a cylindrical wall

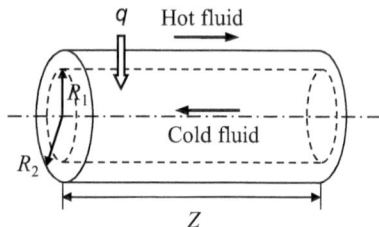

Similar to Eq. (6.20), letting the overall heat transfer coefficient referenced to heat transfer area, A, be U, the total amount of heat, Q [J/s], that transfers from the hot fluid to the cold one can be described by

$$Q = qA = UA (T_h - T_c) \tag{6.23}$$

In Fig. 6.14, the area is different on the inside and outside of the circular tube; thus the value of U varies depending on the area used as basis for calculation. However, Q is constant and independent of the heat transfer surface. In other words, by treating UA collectively as a single term instead of two separate ones, the value becomes constant regardless of the reference surface and direction of heat transfer. Rewriting Eq. (6.22) as expressions of the relationship between driving force (temperature difference) and thermal resistances for the outer surface of the circular tube, the solid wall, and the inner surface of the circular tube, we obtain

$$Q = \frac{T_h - T_{i1}}{2\pi R_2 Z h_h} = \frac{T_{i1} - T_{i2}}{2\pi Z k / \ln (R_2/R_1)} = \frac{T_{i2} - T_c}{2\pi R_1 Z h_c}$$

$$= \frac{2\pi Z (T_h - T_c)}{1/R_2 h_h + \ln (R_2/R_1) /k + 1/R_1 h_c} \tag{6.24}$$

From Eqs. (6.23 and 6.24), UA can be given by

$$UA = \frac{2\pi Z}{1/R_2 h_h + \ln (R_2/R_1) /k + 1/R_1 h_c} \tag{6.25}$$

The inner surface area and the outer surface area of the circular tube are expressed by $2\pi Z R_1$ and $2\pi Z R_2$, respectively. Plugging these terms for A into Eq. (6.25), the overall heat transfer coefficients based on the inside tube heat transfer area, U_1, and that based on the outside tube heat transfer area, U_2, are given by

$$\frac{1}{U_1} = \frac{R_1}{R_2 h_h} + \frac{R_1 \ln (R_2/R_1)}{k} + \frac{1}{h_c}, \quad \frac{1}{U_2} = \frac{1}{h_h} + \frac{R_2 \ln (R_2/R_1)}{k} + \frac{R_2}{R_1 h_c} \tag{6.26}$$

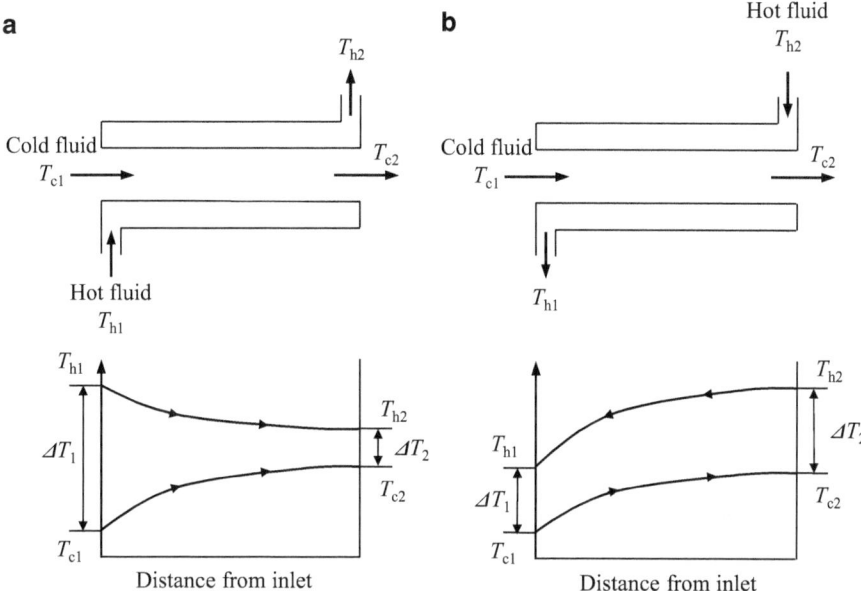

Fig. 6.15 Temperature gradients across (**a**) cocurrent and (**b**) countercurrent double-pipe heat exchangers

(2) *Logarithmic mean temperature difference and calculation for design of heat exchanger*: We need to know the heat transfer area, A, in order to determine the size of a heat exchanger, which is a parameter that can be computed from the total amount of heat transferred, Q; the overall heat transfer coefficient, U (or UA); and temperature difference, $T_h - T_c$, as described by Eqs. (6.20 and 6.23). Note that the temperature difference between the hot and cold fluids, ΔT ($= T_h - T_c$), varies in the flow direction. Figure 6.15 shows typical graphical representations of the temperature changes in the flow direction of the hot and cold fluids (T_h and T_c, respectively) as these fluids flow separately through the annular space and the inner tube of a double-pipe heat exchanger. The flow of the hot and cold fluids in the same direction is termed a *cocurrent flow*, while that in opposite directions to each other is called a *countercurrent flow*. The cold fluid experiences an increase in temperature from the inlet through to the outlet in either case; the hot fluid undergoes a temperature decrease. The temperature difference between the fluids, ΔT, differs by location in the heat exchanger (Fig. 6.15a). In the case of cocurrent flow, the highest ΔT is observed at the inlet, while it decreases toward the outlet, wherein ΔT varies greatly in the flow direction. On the other hand, changes in ΔT with position in heat exchanger in the case of countercurrent flow are not as large as that in the cocurrent flow case. In the condition where the temperature difference, $T_h - T_c$, differs by position

in the above-described manner, the total amount of heat transferred, Q, for the heat exchanger is given by (Fig. 6.15)

$$Q = \frac{UA\,(\Delta T_1 - \Delta T_2)}{\ln\,(\Delta T_1/\Delta T_2)} = UA\Delta T_{lm} \tag{6.27}$$

ΔT_{lm} of Eq. (6.27) represents the logarithmic mean temperature difference which can be determined by

$$\Delta T_{lm} = \frac{\Delta T_1 - \Delta T_2}{\ln\,(\Delta T_1/\Delta T_2)} \tag{6.28}$$

where $\Delta T_1 = T_{h1} - T_{c1}$ and $\Delta T_2 = T_{h2} - T_{c2}$. Refer to Example 6.5 for derivation of Eq. (6.27).

The heat transfer area can be calculated following the steps below for a heat exchanger from the above equations:

- The total amount of heat transferred, Q, for the heat transfer process depicted in Fig. 6.16 can be determined by the following equation:

$$Q = \pm c_{ph} W_h\,(T_{h1} - T_{h2}) = c_{pc} W_c\,(T_{c2} - T_{c1})$$

$$(+ : \text{cocurrent}; -- : \text{countercurrent}) \tag{6.29}$$

where c_{ph} [J/(kg·K)] and c_{pc} [J/(kg·K)] are the specific heat capacities for the hot and cold fluids, respectively, while W_h [kg/s] and W_c [kg/s] are the respective mass flow rates of the hot and cold fluids.
- Calculate the overall heat transfer coefficient, U, from Eq. (6.21) for cases in which the solid wall is a flat plate.
- Determine the logarithmic mean temperature difference, ΔT_{lm}, by Eq. (6.28).
- Substitute the determined values of Q, U, and ΔT_{lm} into Eq. (6.27) to calculate the heat transfer area, A.

Fig. 6.16 Heat transfer across a double-pipe heat exchanger and logarithmic mean temperature difference

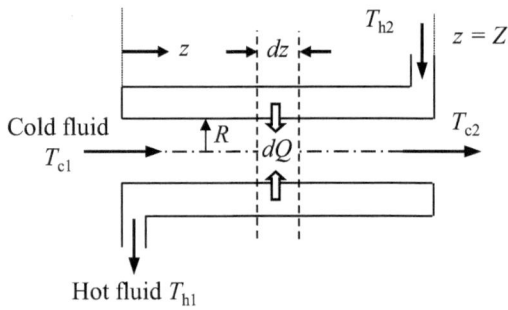

Further, for tubular heat exchanger in which the solid wall is a circular tube, plug the determined values of Q and ΔT_{lm} into Eq. (6.27) to determine UA, and then calculate the length of tube, Z, from Eq. (6.25). Finally compute the heat transfer area by $A_i = 2\pi R_i Z$ ($i = 1, 2$), if necessary.

Example 6.5 Using a cocurrent double-pipe heat exchanger (Fig. 6.16) as an example, demonstrate that the total amount of heat transferred between the hot and cold fluids is described by Eq. (6.27).

Solution Set the inlet of the cold fluid as the origin, use the flow direction as the z-axis, and let the length of the heat exchanger be Z. Consider the heat balance of each fluid for an infinitely small interval on the z-axis, dz. Letting the amount of heat transferred from the hot fluid to the cold fluid within the infinitesimal interval, dz, be dQ [J/s], the temperature decrease in the hot fluid due to the heat transfer be dT_h, and the corresponding temperature increase in the cold fluid be dT_c, the relationship of these terms can be described by

$$dQ = -c_{ph}W_h dT_h = c_{pc}W_c dT_c \tag{6.30}$$

The hot fluid loses heat of dQ, and at the same time, transfer of the exact amount of heat (dQ) occurs over to the cold fluid driven by the temperature difference ($T_h - T_c$). Therefore, letting the radius of the inner pipe be R and the overall heat transfer coefficient be U and knowing that the heat transfer area is given by $2\pi R dz$, Eq. (6.23) can be rewritten as

$$dQ = U (2\pi R dz) (T_h - T_c) \tag{6.31}$$

The following two equations can then be derived from Eqs. (6.30 and 6.31):

$$-\frac{dT_h}{T_h - T_c} = \frac{U2\pi R}{C_{ph}W_h} dz \tag{6.32}$$

$$\frac{dT_c}{T_h - T_c} = \frac{U2\pi R}{C_{pc}W_c} dz \tag{6.33}$$

Adding the two equations above vertically,

$$\frac{d (T_h - T_c)}{T_h - T_c} = -U2\pi R \left(\frac{1}{c_{ph}W_h} + \frac{1}{c_{pc}W_c} \right) dz \tag{6.34}$$

Calculating $1/c_{ph}W_h$ and $1/c_{pc}W_c$ by Eq. (6.29) and then substituting the values in Eq. (6.34) give

$$\frac{d (T_h - T_c)}{T_h - T_c} = -U2\pi R \left(\frac{T_{h1} - T_{h2}}{Q_h} + \frac{T_{c2} - T_{c1}}{Q_h} \right) dz = -\frac{U2\pi R}{Q}$$
$$\times ((T_{h1} - T_{c1}) - (T_{h2} - T_{c2})) dz \tag{6.35}$$

Integrating the left side of Eq. (6.35) from $T_{h1} - T_{c1} = \Delta T_1$ to $T_{h2} - T_{c2} = \Delta T_2$ and the rightmost of the equation with respect to z from 0 to Z, we obtain

$$\ln \frac{T_{h1} - T_{c1}}{T_{h2} - T_{c2}} = \ln \frac{\Delta T_1}{\Delta T_2} = \frac{U 2\pi R Z}{Q} (\Delta T_1 - \Delta T_2) \tag{6.36}$$

Knowing that the heat transfer area is given by $A = 2\pi R Z$, the total amount of heat transferred, Q, is thus derived as

$$Q = \frac{UA\,(\Delta T_1 - \Delta T_2)}{\ln\,(\Delta T_1/\Delta T_2)} = UA(\Delta T)_{\mathrm{lm}} \tag{6.27}$$

◢

Example 6.6 An aqueous sucrose solution is heated from 50 to 70 °C while flowing at a volume flow rate of 40 L/min through a circular pipe (thermal conductivity of the pipe $= 50$ kW/(m · K)) that measures 0.023 m and 0.033 m in inner diameter and outer diameter, respectively. Hot water enters the annular space at 95 °C, flowing on the outside of the pipe countercurrent to the sucrose solution, and discharges through the outlet at 80 °C. The respective heat transfer coefficients of the thermal boundary layers inside and outside the pipe are 2650 W/(m² · K) and 3050 W/(m² · K). Determine the flow rate of hot water, W_h, and the length of pipe, Z, necessary for this heat exchange operation. Let the respective densities of the aqueous sucrose solution and hot water be 1200 and 1000 kg/m³ and the respective specific heat capacities be 3120 J/(kg · K) and 4180 J/(kg · K).

Solution By Eq. (6.29),

$$Q = (3120)\,(0.04 \times 1200/60)\,(70 - 50) = (4180)\,(95 - 80)\,W_h$$

$$Q = 49920 = 62700 W_h \tag{6.37}$$

Hence, $W_h = 0.796$ kg/s. Next, by Eq. (6.25),

$$UA = \frac{2\pi Z}{\dfrac{1}{(2650)(0.0115)} + \dfrac{1}{50}\ln\dfrac{0.0165}{0.0115} + \dfrac{1}{(3050)(0.0165)}} = 104.89Z \tag{6.38}$$

In addition, by Eq. (6.28),

$$\Delta T_{\mathrm{lm}} = \frac{30 - 25}{\ln\,(30/25)} = 27.42 \tag{6.39}$$

Hence, substituting Q, UA (from Eq. (6.38)), and $(\Delta T_c)_{\mathrm{lm}}$ (from Eq. 6.39) into Eq. (6.27) gives

$$Z = \frac{49920}{(104.89)(27.42)} = 17.35 \text{ m} ◢$$

6.3 Low-Temperature Storage of Produce

6.3.1 Reasons for Low-Temperature Storage of Fresh Foods

Deterioration in terms of quality (decrease in quality) occurs with time when fresh foods such as vegetables, fruits, fishes, and meats are left to stand at room temperature over an extended period of time. In particular, freshly harvested vegetables, fishes, and meats are foods that require special attention. There are several possible causes of food degradation: ① even after harvest, fruits and vegetables consisting of cell tissues still keep on breathing and metabolizing, leading to aging and degradation; ② enzymatic degeneration of cells by certain types of indigenous enzymes in fishes, meats, etc., comes into play after death of the animals; and ③ parasitized by microorganisms, food compounds are decomposed, causing spoilage.

Respiratory metabolism of fruits and vegetables, enzymatic destruction of cellular tissues, microbial proliferation are all biochemical reactions. The reaction rate constants of these reactions exhibit Arrhenius-type temperature dependence described in Sect. 6.2.2, wherein the rate constants vary exponentially with temperature (increase with rising temperature). Therefore, the degradation in quality can be delayed by storage at low temperatures in order to lower the reaction rate. The storage life of foods is said to be extended by two- to threefolds by reducing the storage temperature by 10 °C (K) through refrigeration.

Fruits and vegetables contain significant amounts of water which has a crucial influence on quality deterioration of the food. Water provides a ground for microbial proliferation and degradation reactions. In order to make long-term storage possible, it is important to either eliminate water by drying operations as described in Sect. 4.4 or by freezing the water making it unavailable as a reaction medium as described hereafter. Water present in food materials does not freeze at 0 °C due to freezing-point depression; the water will not freeze unless the food materials are being stored at temperatures below the freezing point. Moreover, fresh foods are stored refrigerated as a means of preservation at low temperatures ranging from −1 to 4 °C within which water stays unfrozen to avoid the physical influences associated with ice formation (e.g., concentration of solution). In contrast to refrigeration, frozen storage involves cooling of a food product to low temperatures such that just about all the water contained in the food is frozen into ice crystals and, in addition, the need to remove the latent heat of freezing released during ice formation. Depending on the final storage temperature, preservation of up to about 1 year may be possible at −30 °C. In the production of frozen foods, food materials are generally subjected to blanching with boiling water or the like for inactivating the enzymes contained within the food materials prior to freezing.

6.3.2 Refrigeration and Freezing Systems

Refrigeration of food is an operation of removing the sensible heat from the food in question, while freezing refers to further cooling to also remove the latent heat

Fig. 6.17 Schematic diagram
of a typical refrigeration and
freezing system

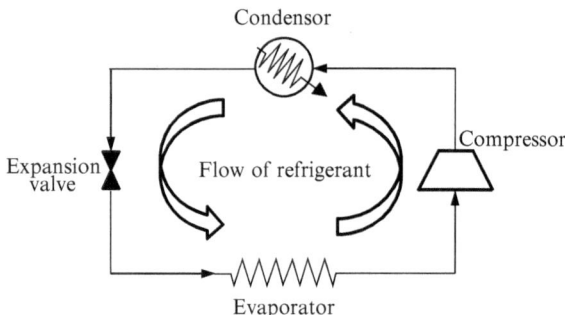

of freezing of the water contained within to generate ice crystals. Both operations involve removal of heat from food materials, which are a kind of heat exchange operation like the one discussed in Sect. 6.2.4. In the old days, refrigeration of fresh foods was performed using ice; in recent years, the old method has been increasingly replaced by means of refrigeration and freezing systems which use liquid refrigerants, such as ammonia and Freon, and are able to perform both refrigeration and freezing functions. Figure 6.17 pictures schematically a typical piece of refrigeration and freezing equipment that consists of a compressor, a condenser, an expansion valve, and an evaporator. When the compressed refrigerant in the liquid state passes through the expansion valve, part of the refrigerant vaporizes, resulting in a liquid and vapor mixture of the refrigerant. As this mixture refrigerant passes through the evaporator, the liquid part of the refrigerant vaporizes, depriving the surroundings of the evaporator of heat. Here, all the refrigerant is transformed into vapor. Subsequently, the refrigerant vapor is compressed by the compressor and transforms back to the liquid state in the condenser. As illustrated in Fig. 6.17, a typical refrigeration cycle removes heat energy from the surroundings through repetition of the compression and evaporation processes of refrigerant powered by the energy supplied to the compressor.

Refrigeration and freezing equipment is classified based on the contact method between the food products and the evaporator which is the heat exchange part with the surroundings. For liquid foods such as milk, scraped-surface heat exchangers are also used as evaporator in addition to plate heat exchangers, shell and tube heat exchangers, etc. Figure 6.18 shows a couple of examples of refrigeration and freezing units for solid fresh foods. The one pictured in Fig. 6.18a is a unit that uses air as cooling medium. The cold air obtained via exposure to the evaporator is brought into contact with the food materials placed on a spiral conveyor in order to refrigerate or freeze the food. Figure 6.18b shows a system that refrigerates or freezes the food on a belt conveyor by blowing cold air onto the food through nozzles. The system is designed such that cold air runs through the back of the food materials by making use of the Coanda effect of cold air. The Coanda effect refers to a phenomenon that arises when a fluid jet entrains its surrounding fluid due to viscous effect. Fruits and vegetables are often refrigerated by direct contact

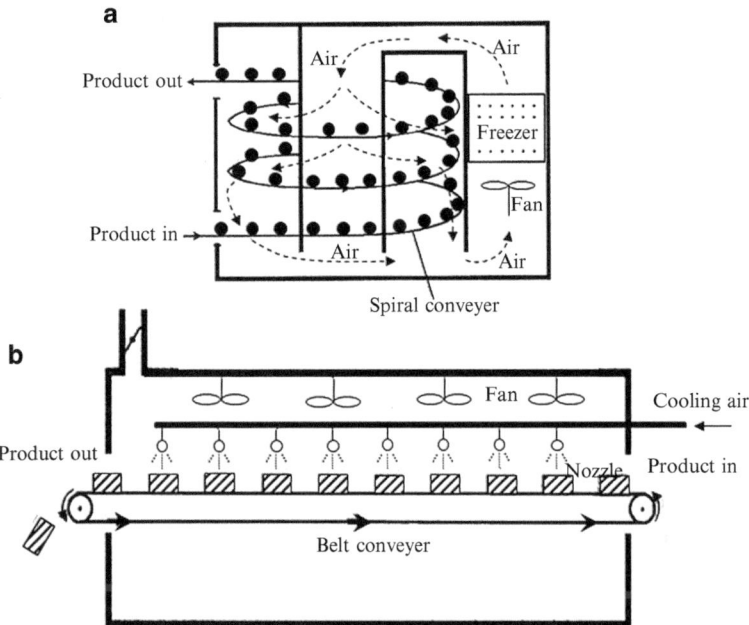

Fig. 6.18 Refrigeration and freezing units for solid fresh foods

with water chilled beforehand with a cooling machine. Such operation using chilled water is performed to combine refrigeration with cleaning of harvested produce.

6.3.3   Refrigeration and Refrigeration Time of Fruits/Vegetables

(1) *Calculation of the amount of heat to be removed*: Refrigeration refers generally to low-temperature preservation method of maintaining the temperature within the range of −1 to 4 °C, by which sensible heat and metabolic heat from the target fresh foods are removed. The total amount of heat, Q_T [J], that must be removed when cooling a target food from temperature T_1 to T_2 can be calculated by

$$Q_T = M \int_{T_1}^{T_2} c_p dT \qquad (6.40)$$

where M [kg] denotes the mass of the food and c_p [J/(kg·K)] denotes the specific heat capacity. When c_p [J/(kg·K)] is a constant value independent of temperature, Eq. (6.40) can be rewritten as

$$Q_T = Mc_p \int_{T_1}^{T_2} dT = Mc_p (T_2 - T_1) \qquad (6.41)$$

Fruits and vegetables contain comparatively more water than other food. While the specific heat capacity of water is 4.18 kJ/(kg·K), those of carbohydrates, proteins, and lipids are approximately 1.5 kJ/(kg·K) which is less than half of that of water. Considering food materials as mixtures of water and other compounds and assuming additive property of specific heat capacity, the specific heat capacity, c_p [J/(kg·K)], of a food material is usually calculated by the following equation:

$$c_p = wc_{pw} + (1 - w) c_{ps} \qquad (6.42)$$

where w [kg-water/kg] represents wet basis moisture content and c_{pw} [J/(kg·K)] and c_{ps} [J/(kg·K)] denote the respective specific heat capacities of water and the drying material.

(2) *Calculation of refrigeration time using the Gurney-Lurie charts*: Here, let us turn to solid foods. The operation of cooling a food from room temperature to lower temperatures can be considered as an unsteady-state heat conduction in which the temperature of the food material changes with time. Therefore, the temperature change in food material can be strictly determined by solving mathematically equations for unsteady-state heat conduction. Heat that transfers by conduction from the inside of food material to the surface is removed to the surrounding fluid (usually air) by convection. When a food material is brought into direct contact with a cooling plate, the surface temperature of the food material is equal to the temperature of the cooling plate, at which it is kept constant.

Temporal change in temperature of food materials can be determined using the Gurney-Lurie charts (Fig. 4.7a–c) introduced in the discussion on extraction rate in Sect. 4.2.3. We shall heed the following points when using Fig. 4.7:

(i) Letting the initial temperature (constant) of a food material be T_0, the ambient temperature be T_1, and the temperature at certain location inside the food material at time t be T, Y that goes on the ordinate of Fig. 4.7 is defined as follows:

$$Y = \frac{T_1 - T}{T_1 - T_0} \qquad (6.43)$$

(ii) Temperature conductivity, $\alpha = k/(\rho \cdot c_p)$, goes on the abscissa of Fig. 4.7 in place of diffusion coefficient, D, where k [W/(m·K)] denotes the thermal conductivity of the food material, ρ[kg/m³] the density, and c_p [J/(kg·K)] the specific heat capacity.

(iii) When using the Gurney-Lurie charts that take into account the heat transfer coefficient of the surface of food material, $m = k/(h·R)$ shall be used instead of $m = D/(k_m \cdot R)$, where h represents the heat transfer coefficient of the thermal boundary layer at the surface of food material and R denotes half the thickness in the case of a flat plate or the radius in the case of a cylinder or a sphere. In cases where the heat transfer coefficient, h, is extremely large, m becomes 0, and the surface temperature of food material in Eq. (6.43) becomes equal to the ambient temperature, T_1.

Example 6.7 A melon 15 cm in diameter (thermal conductivity $= 0.56$ W/(m $\cdot$ K), density $= 930$ kg/m^3, and specific heat capacity $= 3.8$ kJ/(kg $\cdot$ K)) is stored in a 4 °C refrigerator starting with an initial temperature of 30 °C. How long does it take for the temperature at the center of that melon to reach 10 °C? Let the heat transfer coefficient between the melon and the surrounding atmosphere be 15 W/(m^2 $\cdot$ K).

Solution We first calculate the parameters necessary for the determination of $\alpha t/R^2$ using the Gurney-Lurie chart. $R = 0.075$ m, $\alpha = k/(\rho \cdot c_p) = 1.58 \times 10^{-7}$ m^2/s, and $m = k/(h \cdot R) = 0.56/[(15)\ (0.075)] = 0.5$, and because we are determining the temperature at the center, $n = 0$. When the center temperature of melon, T_c, reaches 10 °C, the value of Y will be equal to $(T_1 - T_c)/(T_1 - T_0) = (277 - 283)/(277 - 303) = 0.231$. Reading the x-coordinate of the straight line of $m = 0.5$ and $n = 0$ at $Y = 0.231$ from Fig. 4.7a, $\alpha t/R_2$ is determined to be 0.446. Hence, the center temperature reaches 10 °C after a lapse of $t = (0.446)\ (0.075)^2/(1.58 \times 10^{-7}) = 16{,}082$ s $= 4.47$ h. ◢

6.3.4 Freezing and Thawing

(1) *Blanching*: Blanching is a pretreatment fundamental in the freezing process of vegetables. If a vegetable is stored at -20 °C directly after harvest for several weeks, alterations in terms of flavor, color, and texture will occur. Significant deterioration in quality follows as storage period is further prolonged. The above-described quality changes are presumably attributed to the action of enzymes which are not inactivated by freezing. Peroxidase and catalase in fruits and vegetables have high cold tolerance and are known causes of quality deterioration at low temperatures. In the late 1920s, short-time heat treatment technology for deactivating the enzymes in vegetables prior to freezing has been developed, and the operation was named blanching. Blanching not only inactivates enzymes but also exercises other combined effects such as removal of the oxygen contained in vegetables, reduction in number of microorganism cells, and improvement of texture. However, improper conditions of heat treatment may end up accelerating quality deterioration by causing losses of vitamin C and aromatic essential oils. Blanching refers to the short heat treatment of bringing vegetables and fruits into contact with hot water and steam. Among the various methods, blanching using steam (or superheated steam) is able to produce finished products of lower water contents and in addition with less loss of water-soluble components contained in the vegetables.

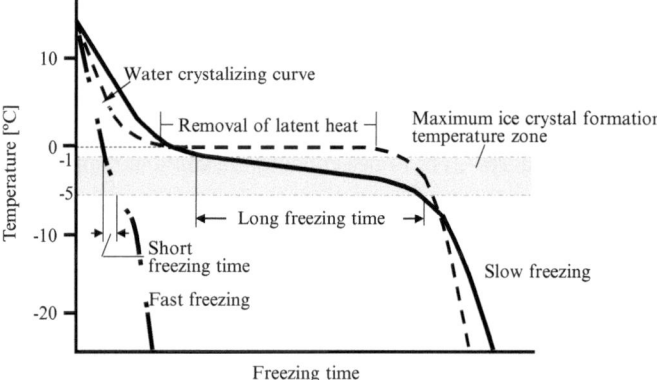

Fig. 6.19 Maximum ice crystal formation temperature zone and mechanisms of freezing

Besides, blanching techniques that require even shorter time using microwaves have also been developed recently.

(2) *Freezing and thawing mechanisms*: The effects produced by the ice crystals resulted from freezing pose a challenge in frozen storage of foods. The size of the ice crystals in particular has noticeable influences on the food quality after thawing. Hence, it is of great importance to understand the freezing and thawing mechanisms.

Vegetables and fruits are approximately 80–90 % water. This water turns into ice when it is cooled. During the freezing process of food, the sensible heat of the food material is first removed and then followed by the latent heat of freezing. As shown in Fig. 6.19, pure water freezes into ice crystals at 0 °C, and the temperature remains constant throughout the entire freezing process. However, in actual food materials, formation of ice crystals only commences at temperatures below 0 °C because of freezing-point depression induced by the various substances such as amino acids, sugars, and salts that dissolve in the water. Furthermore, freezing of food materials is a phenomenon that takes place within a certain temperature range instead of happening instantaneously at a particular temperature. In general, roughly 80 % of the water contained in food materials is believed to freeze at a temperature range of −1 to −5 °C, which is known as the zone of maximum ice crystal formation. In this temperature zone, formation of ice results in the release of latent heat of freezing, making the food temperature comparatively less likely to decrease than that when ice formation does not occur and thus taking the food material longer to pass through the zone. The rate at which a food material passes through the temperature zone (freezing rate) dictates the size of the ice crystals. Slow freezing results in the formation of larger ice crystals; rapid freezing generates fine ice crystals. During slow freezing, large ice crystals form in the gap between cells. Volume expansion and rise in intracellular osmotic pressure arising from the formation of large ice crystals cause not only cellular damage but also outflow of useful

Fig. 6.20 Comparison between freezing and thawing processes

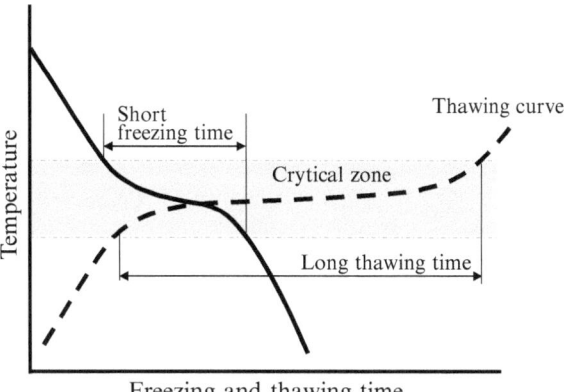

Freezing and thawing time

compounds from the inside to the outside of the ruptured cells. The slow-frozen food loses water known as drip along with many other useful compounds upon thawing, resulting in detrimental quality changes. On the contrary, food materials pass through the temperature zone of maximum ice crystal formation in a much shorter time during rapid freezing as shown in Fig. 6.19, and the water inside and outside the cells is dispersed and frozen instantaneously in a fine amorphous state. Therefore, the cells are not damaged and the quality of food is preserved.

Thawing operation also has a significant impact on food quality. Figure 6.20 shows schematically the temperature changes during freezing and thawing processes. Thawing is not simply the reverse of freezing, and it can take far longer time than that required by freezing because the thermal conductivity of water (0.58 W/m · K) is smaller than that of ice (2.2 W/m · K). Since foods in the frozen state are mixtures of ice and food components as opposed to the ones of water and food components in the unfrozen state, the apparent thermal conductivity of frozen foods is thus comparatively higher. During freezing and thawing, frozen and unfrozen layers form adjacent to the outer surfaces of food materials, across which heat transfer occurs by conduction. Therefore, given the same temperature difference between the inside and outside of a food material, the heat transfer rate through the food during freezing can be more than three times of that through the exact same food during thawing (see Example 6.7). It is this difference in the rate of heat transfer that causes the time required to thaw a food material to be longer than that required to freeze it.

(3) *Calculations of freezing, thawing times and the amount of heat to be transferred*: Freezing involves an unsteady-state heat transfer which is further complicated by the removal of the latent heat of freezing over a range of temperatures. Therefore, the Gurney-Lurie charts are not applicable for the determination of freezing time. It is exceptionally complex to precisely calculate the freezing time due to the requirement to take into consideration the changes in the thermal properties (thermal conductivity, etc.) of the

Fig. 6.21 Plank's model for
estimating freezing time

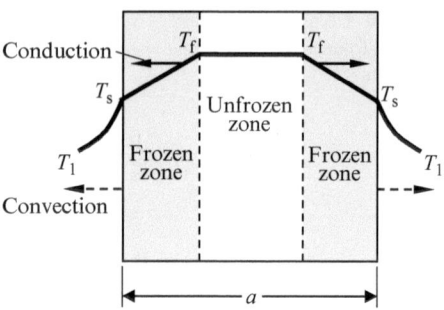

material during freezing. Plank's equation developed based on the following
assumptions is commonly used for approximate estimation of freezing time:

(i) The food material is uniform in terms of initial temperature at its freezing
point with all part of the food still unfrozen (without formation of ice
crystals).
(ii) The thermal conductivity of the frozen food material is constant.
(iii) The heat transfer in the frozen layer of the food material approximates
steady-state heat conduction.

Imagine that a food slab of thickness a is being placed in an atmosphere
of ambient temperature T_1 in order to freeze it from two sides of its vertical
surfaces (Fig. 6.21). Let the freezing point be T_f [°C] and the heat transfer
coefficient between the food slab and the surrounding fluid (usually air) be h
[W/(m$^2 \cdot$ K)]. Heat transfer takes place by conduction in the frozen layer that
forms at the surfaces of the food slab as freezing progresses and the temperature
distribution established in the frozen layer is assumed to be linear. Letting
the thermal conductivity of the frozen layer be k [W/(m $\cdot$ K)], its density be
ρ [kg/m^3], the latent heat of freezing per unit mass for the unfrozen phase be
ΔH_f [J/kg], the time required to completely freeze the food slab, t [s], is thus
given by

$$t = \frac{\Delta H_f \rho}{T_f - T_1} \left(\frac{a}{2h} + \frac{a^2}{8k} \right)$$

(6.44)

Equation (6.44) is generalized as follows when the shape of the food material
approximates a sphere or an infinite cylinder:

$$t = \frac{\Delta H_f \rho}{K (T_f - T_1)} \left(\frac{a}{2h} + \frac{a^2}{8k} \right)$$

(6.45)

For an infinite slab, the parameter a in Eq. (6.45) represents the thickness of
the slab; for an infinite cylinder or a sphere, it represents the diameter. K takes
the value of 1 for a slab, 2 for a cylinder, or 3 for a sphere. Further, in a food

material that has a wet basis moisture content of w, the term ΔH_f in Eqs. (6.44 and 6.45) thus becomes $\Delta H_f \cdot w$.

Equation (6.45) was developed on the basis of the assumptions that the initial temperature of the food material is equal to the freezing point and the temperature is uniform throughout the food material. In the case where the initial temperature of the food material differs from the freezing point, the latent heat of freezing corrected for sensible heat, $\Delta H_f'$, obtainable by the following equation is used in place of ΔH_f as the latent heat of freezing:

$$\Delta H_f' = \Delta H_f + c_{pu} (T_i - T_f) \qquad (6.46)$$

where c_{pu} [W/(kg · K)] is the specific heat capacity of the unfrozen phase and T_i [°C] is the initial temperature of the food material.

From a heat transfer point of view, since thawing is the reverse phenomenon of freezing, the concept based on which Eq. (6.45) for estimation of freezing time is derived is directly applicable to estimation of thawing time. In thawing, since the ambient temperature is higher than the freezing point of the food material and heat transfer occurs in the direction opposite to that in a freezing process, the temperature difference, $T_f - T_1$, of Eq. (6.45) has to be replaced with $T_1 - T_f$. Besides, both the thermal conductivity and density of the melted phase (unfrozen phase) shall be used.

The total amount of heat, Q_T [J], removed when freezing a food material of mass M [kg] and wet basis moisture content w [kg-water/kg] are expressed by the following equation derived by adding the latent heat of freezing to Eq. (6.41) which expresses the load of sensible heat for a refrigeration operation:

$$Q_T = M \left[c_{pu} (T_i - T_f) + w\Delta H_f + c_{pf} (T_f - T_p) \right] \qquad (6.47)$$

where T_i is the initial temperature of the food material, T_f is the freezing point, T_p is the final temperature, and ΔH_f is the latent heat of freezing. c_{pu} [J/(kg · K)] and c_{pf} [J/(kg · K)] denote the specific heat of the unfrozen and frozen food materials, respectively, which can be calculated by plugging in the wet basis moisture content or ice content for w and the specific heat of water or ice for c_{pw} in Eq. (6.42).

Example 6.8 Water of 0 °C is filled into a stainless steel container with a cross-sectional area of 1 m × 15 cm and measuring 1.5 m in depth. The container is then immersed into a cooling medium of −18 °C in order to freeze the water. Assume that heat transfer (removal of heat energy) occurs dominantly through the two surfaces of 1 m × 1.5 m, while heat transfer across other surfaces is negligible. Calculate the freezing time under this condition using Plank's equation for infinite slab (Eq. 6.44). Then, determine using the same equation the thawing time required when the ice of 0 °C in the same container is thawed (to transform it back to water of 0 °C) by immersing the container into a heating medium of 18 °C. Assume that convection does not occur in the water inside the container during the freezing process, that the heat transfer coefficient of the outer surface of the container is substantially large, and that the heat resistance of stainless steel is negligible. Let the respective densities of water and ice be 1000 and 917 kg/m^3 and their respective thermal conductivities be 0.58 and 2.22 W/(m · K).

Solution The freezing material is a slab, so $K = 1$. We may assume $h \to \infty$ since the cooling medium is well agitated. Plugging in 3.34×10^5 J/kg for the latent heat of freezing of water, ΔH_f, 1000 kg/m^3 for the density of water, 2.22 W/(m · K) for the thermal conductivity of water, 0 °C for T_f, and -18 °C for T_1 in Eq. (6.44), the freezing time, t_1, is determined to be

$$t_1 = \frac{\Delta H_f \rho}{T_f - T_1} \left(\frac{a}{2h} + \frac{a^2}{8k} \right) = \frac{(1000)\,(3.34 \times 10^5)}{18} \frac{0.075^2}{(8)(2.22)} = 5876 \text{ s} = 1.63 \text{ h}$$

On the other hand, when Eq. (6.44) is applied to the thawing process, the thawing time, t_2, can be determined by

$$t = \frac{\Delta H_f \rho}{T_1 - T_f} \left(\frac{a}{2h} + \frac{a^2}{8k} \right)$$

where T_1 is the temperature of the heating medium used in the thawing process. Hence,

$$t_2 = \frac{(1000)\,(3.34 \times 10^5)}{18} \frac{0.075^2}{(8)(0.583)} = 22380 \text{ s} = 6.22 \text{ h}$$

There is a large difference between the freezing time and the thawing time. This difference is attributable to the difference in heat transfer properties (thermal conductivity) between the frozen phase and the melted phase. ◢

6.4 Jam

6.4.1 Manufacturing Process of Jam

Jams generally refer to concentrated mixtures of fruit pulp or juice and sugar, the composition of which is controlled by law in some countries. While some jams may retain the form of fruit pulp (preserved style), some may not. During the manufacturing process, *pectin* is contained in the fruits gels through the action of organic acids in combination with the high sugar concentration. Pectin is a complex polysaccharide composed mainly of galacturonic acid residues which are partially esterified by methyl groups, resulting in pectin also carrying methoxyl groups. Pectin molecules in fruits mostly contain 7 % or more methoxyl groups and are classified as high-methoxyl or HM pectin which shows gelation behaviors when

Table 6.4 Sugar contents, acidities, and pectin contents of raw fruits for producing jams

Fruit	Sugar content (Brix)	Acidity (%)	Pectin content (%)	Sour taste (major organic acid)
Apple	10 ~ 15	0.5 ~ 1.0	ca. 0.6	Malic acid
Apricot	7 ~ 8	1.2 ~ 2.3	ca. 0.8	Malic acid, citric acid
Fig	7 ~ 10	ca. 0.3	ca. 0.7	Malic acid
Grape	12 ~ 16	0.6 ~ 1.0	0.2 ~ 0.3	Tartaric acid
Peach	9 ~ 19	0.3 ~ 0.6	ca. 0.6	Malic acid, citric acid
Prune	ca. 15	ca. 1.2	ca. 0.7	Malic acid
Raspberry	11 ~ 13	0.4 ~ 1.0	1.0 ~ 2.1	Citric acid
Strawberry	5 ~ 11	0.5 ~ 1.0	ca. 0.6	Citric acid

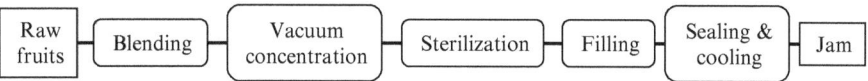

Fig. 6.22 Manufacturing process of jam

heated in the presence of sugar and organic acids. Carboxyl groups, the dissociation of which is inhibited in the presence of organic acids, present in pectin molecules facilitate the formation of hydrogen bonds, thus imparting firmness to the gel.

Table 6.4 summarizes the sugar content, acidity, pectin content, and main organic acids that give a jam its sour taste contained in fruits as raw materials of jams. The sugar content that represents the quantity of water-soluble solids can be determined with a refractometer. The acidity is determined by titration with sodium hydroxide and expressed in % (w/w) as citric acid equivalent. Sugar holds the network structure of pectin together, promotes hydrogen bonding, and thus stabilizes the gel. This effect usually manifests itself at a sugar concentration of 50 % or more, but a concentration of 60 % or above will be necessary for creating a gel of good texture. Because the sugar contained in the fruits alone is normally insufficient, additional sugar is added during the manufacturing process. At 65 % sugar concentration and above, the water activity (discussed in the successive subsection) drops down to or below 0.85, preventing the growth of yeasts and microbes and hence imparting storage stability to the jams.

Figure 6.22 depicts a main manufacturing process of jam. Fruits—the raw material—are subjected to screening and cleaning before being blended with other raw materials and concentrated at low temperatures using a vacuum concentrator. Then, after high-temperature short-time sterilization, the concentrate is filled immediately into jars, tightly sealed, and cooled, and finally the jam is ready. In some cases, sterilization is performed after tight sealing of the concentrate-filled jars. As shown in Table 6.4, as a raw material of jams, fruits contain only around 10 % sugar which is insufficient for gelation to occur. Sugar is added not only for gelation purposes but also for the enhancement of storage stability of jams by lowering the water activity as described later in the chapter.

Fig. 6.23 Structure of water
molecule

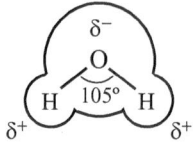

6.4.2 Water Activity and Quality Deterioration of Foods

Water, consisting of one atom of oxygen and two atoms of hydrogen, has a shape
of a distorted tetrahedron with the oxygen atom located at the center and the two
hydrogen atoms located one each at two of the four corners of the tetrahedron.
The two hydrogen atoms are attached to the oxygen atom and separated at the
bond angle of 105° (H-O-H bond angle) (Fig. 6.23). The oxygen atom in a water
molecule has two lone pairs of electrons and thus a partial negative charge (δ^-),
while the hydrogen atom has a partial positive charge (δ^+). As just described,
a substance made up of molecules in which positive and negative charges are
distributed nonuniformly on the various atoms is called a dipole. Due to such
properties, water molecules tend to interact with each other or with other substances
as well through the formation of hydrogen bonds.

Water contained in food materials can be separated into free water that has weak
interactions with other components and can thus move around freely and bound
water that interacts with other components (sugars, proteins, salts, etc.) and is to
some extent limited in mobility. Free water can be used for microbial growth in
food. Besides, it freezes at 0 °C and boils at 100 °C under atmospheric conditions.
On the other hand, bound water is not available for microbial growth, and it has
a freezing point below 0 °C and a boiling point above 100 °C under atmospheric
conditions. Even at the same moisture content, the condition of water may vary
depending on the type and composition of the food material; therefore the amount
of bound water may differ as well.

Water activity is an indicator of the proportion of water contained in a food
material that behaves like free water. When a food is tightly sealed in a container
and let stand at a constant temperature for a long period of time, the water vapor
pressure, p, in the headspace will eventually arrive at a constant value. If all the
water contained in the food was free water, the vapor pressure would have been
equal to that of pure water, p_s, stored under the exact same conditions. However,
since part of the water is present in the form of bound water by interacting with the
constituent components of the food and thus does not evaporate, the water vapor
pressure given by the food, p, is lower than p_s. The ratio of p to p_s is called water
activity, a_w, which is an indicator of the proportion of free water in a food material.

$$a_w = p/p_s \tag{6.48}$$

Example 6.9 Aqueous sucrose solutions of various molal concentrations
(molality), m, were measured for water vapor pressure, p, at 20 °C, and the results

Table 6.5 Vapor pressure of sucrose solution

m [mol/kg]	0.5	1.0	2.0	3.5	5.0
p [kPa]	2.319	2.295	2.245	2.157	2.055
a_w	0.9908	0.9810	0.9592	0.9216	0.8780
x_S	0.00892	0.01768	0.03475	0.05926	0.08256

Fig. 6.24 Water activity of aqueous sucrose solution

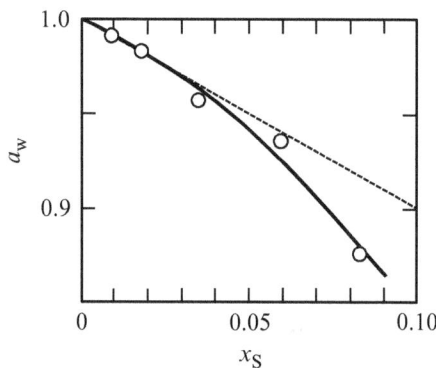

are tabulated in Table 6.5. Determine the respective water activities, a_w, of the aqueous sucrose solutions. The saturated water vapor pressure of pure water, p_s, at 20 °C is 2.34 kPa. Discuss the deviation from Raoult's law by presenting graphically the relationship between the mole fraction of sucrose, x_s, in the aqueous solutions and the water activity.

Solution The third row from top in Table 6.5 shows the water activity value, a_w, calculated according to the definition given by Eq. (6.48). Molal concentration (or molality), m, is defined as the amount of substance of a solute in the unit of mole contained in every kilogram of solvent. In this case, water is the solvent (molar mass = 18 g/mol); hence the amount of substance of 1 kg (1000 g) of water is $1000/18 = 55.56$ mol. The mole fraction of sucrose (solute), x_s, in the aqueous sucrose solutions of molality m is given by

$$x_S = m/(m + 55.56) \qquad (6.49)$$

The mole fraction of sucrose, x_s, obtained by Eq. (6.49) is shown in the fourth row from top of Table 6.5. Also, the relationship between the mole fraction of sucrose, x_s, and the water activity, a_w, is depicted in Fig. 6.24. Raoult's law states that the vapor pressure, p, of a solvent (water) above a solution is proportional to the saturated vapor pressure, p_s, of the pure solvent at the same temperature multiplied by the mole fraction, x_w, of the solvent present in the solution. Since the aqueous sucrose solutions are comprised of the two components of sucrose and water, the relationship of $x_w = 1 - x_s$ holds. Therefore, if Raoult's law is obeyed,

$$a_w = p/p_s = 1 - x_S \qquad (6.50)$$

The straight line (dashed line) in Fig. 6.24 represents the relationship described by Eq. (6.50). The water activity obtained from the vapor pressure of water is below the straight line, and the deviation from the straight line (Raoult's law) grows bigger as sucrose concentration increases. This phenomenon can be interpreted as attributable to the relative reduction in the ratio of free water resulted from the rise in the ratio of water molecules interacting with sucrose molecules as sucrose concentration increases. In addition, greater deviation from the straight line indicates stronger interactions between the solute and the water molecules. ◢

Refer to Example 3.12 on how to determine the water activity of solid food materials. The degree to which a food material retains water is expressed by the moisture sorption isotherm of the food material that depicts the relation between the water activity at a constant temperature and the moisture content (see Chap. 3). A food material generally displays different profiles of moisture sorption isotherms for wetting (sorption of moisture) and drying (desorption of moisture): at an identical water activity, the food material that is drying has relatively higher moisture contents than the exact same food material that is taking up moisture (Fig. 6.25). The afore-described difference in moisture content of a food material between its sorption and desorption isotherms is called the *isotherm hysteresis*, the shape of which varies markedly with different types of foods.

Water activity of food affects the various changes that occur during its storage. Figure 6.26 shows on a conceptual basis the influences water activity has on microbial propagation, enzyme activity, and nonenzymatic browning reaction. Although most microorganisms do not proliferate at a water activity of lower than 0.7, the lower limit of water activity for growth differs by types of microorganisms. In general, fungi propagate at water activities of not lower than 0.7, yeasts not lower than 0.75, and bacteria not lower than 0.85. Therefore, lowering the water activity by addition of sugar like that performed in the preparation of jams makes the environment unsuitable for microbial proliferation and thus enhances the storage stability. Meanwhile, the enzymes present in food materials are still enzymatically active even at a water activity of as low as around 0.4. Furthermore, nonenzymatic browning reaction still progresses down until a water activity of around 0.2.

Fig. 6.25 Moisture sorption isotherms

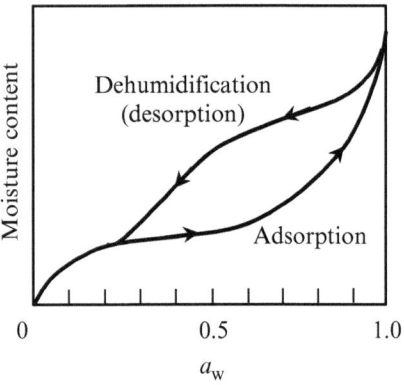

Fig. 6.26 Effects of water activity on chemical and enzymatic reactions and microbial growth

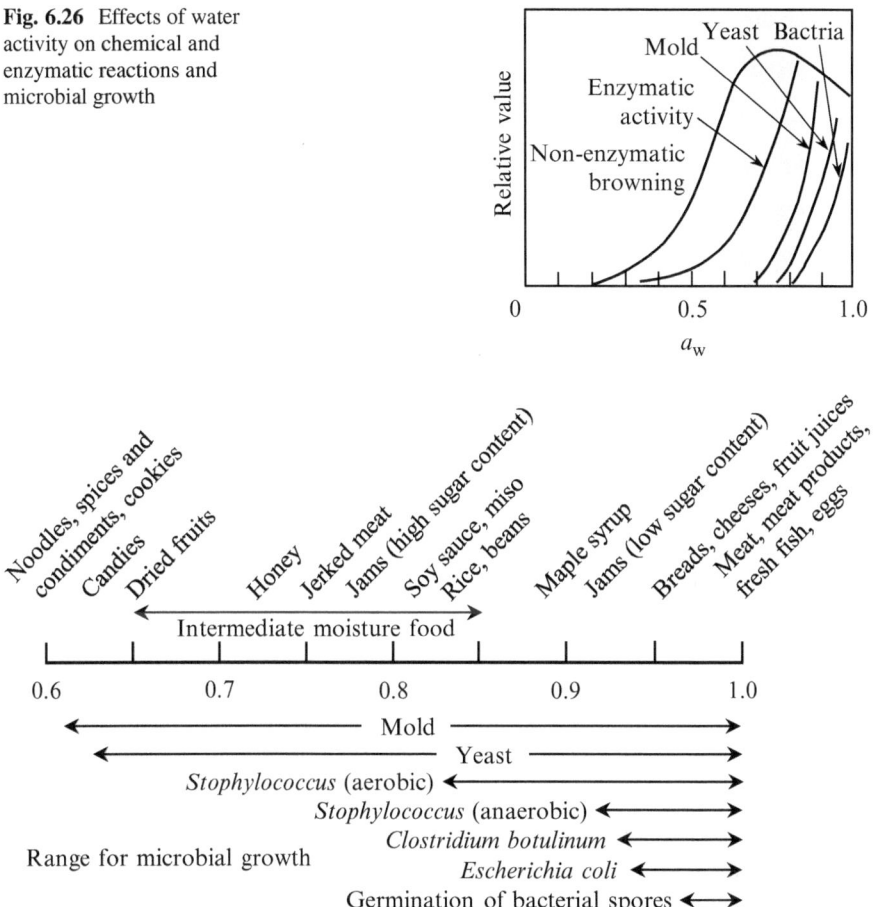

Fig. 6.27 Water activities of some common foods and ranges of water activity for microbial growth

The water activities for some common foods and the ranges of water activity for microbial growth are shown in Fig. 6.27. There are some minor variations in the ranges of water activity for microbial growth between Figs. 6.26 and 6.27. For example, the minimum water activity for fungal growth is 0.7 in Fig. 6.26 as opposed to 0.61 shown in Fig. 6.27. Microorganisms are diverse in kind: common fungi have a minimum water activity for growth of about 0.8, while drought-resistant ones can still grow even at a water activity of approximately 0.65. The difference between Figs. 6.26 and 6.27 stems from the biodiversity of microorganisms; therefore the ranges of water activity for microbial growth shown therein are to be used only as a guide.

6.4.3 Intermediate-Moisture Foods

Foods with water activity values adjusted between 0.65 and 0.85 for improvement of storage stability against microorganisms by adding sugar, salt, etc., as in jams or by drying, dehydration, etc., in order to reduce the moisture content as in jerked meat are known as *intermediate-moisture foods*. Although the moisture content of intermediate-moisture foods has been reduced to some extent for suppression of microbial growth, they still give a moist texture owing to the comparatively higher moisture contents.

There are several methods for manufacturing intermediate-moisture foods:

(1) Partial drying method by partially evaporating water to concentrate the solutes present in the food materials and decreasing the water activity
(2) Wet infusion method (or equilibration method) by soaking solid food materials in an aqueous solution of humectant such as sorbitol and dietary salt to adjust the a_w to a prescribed value
(3) Dry infusion method (or simply known as infusion method) by freeze-drying solid food materials to create a porous structure before soaking it in a solution of humectant with desired a_w
(4) Direct mixing method by blending materials of different a_w and adjusting the a_w of the finished product to a prescribed value.

Exercise

6.1 Table 6.6 shows the heating time, t, and survival rate, N/N_0, during a sterilization process to eliminate a certain type of microorganism at 110 °C. Determine the D-value under this condition.

6.2 The relationship between time, t, and the unoxidized fraction, Y, described by Eq. (6.51) holds for the autoxidation of methyl linoleate:

$$\ln \frac{1-Y}{Y} = kt + \ln \frac{1-Y_0}{Y_0} \tag{6.51}$$

Here, k [h^{-1}] denotes the rate constant. The two parameters of the oxidation process of methyl linoleate at 65 °C are tabulated in Table 6.7. Determine the rate constant, k.

Table 6.6 Thermal death process of a microorganism

Time t [min]	3	6	9	12	15
Survival rate N/N_0	0.351	0.119	0.0431	0.0147	0.00531

Table 6.7 Oxidation of methyl linoleate

Time t [h]	6	9	12	15	18
Unoxidized fraction Y [−]	0.929	0.808	0.611	0.352	0.165

Table 6.8 Temperature dependence of oxidation rate constant, k, of methyl linoleate

Temperature [°C]	30	40	50	60	70
k [h^{-1}]	2.89×10^{-2}	6.32×10^{-2}	1.23×10^{-1}	2.54×10^{-1}	4.68×10^{-1}

6.3 The oxidation rate constants of methyl linoleate are measured at various temperatures (Table 6.8). Determine the activation energy and frequency factor for the autoxidation of methyl linoleate.

6.4 One liter of water of 15 °C is heated to 95 °C in an electric kettle with a 1250-W heating element. The kettle weighs 0.7 kg and has an average specific heat of 0.75 kJ/(kg · °C). Let the specific heat of water be 4.18 kJ/(kg · °C). (a) Determine the time required for heating assuming that there is no heat loss from the kettle. (b) Letting the electricity rate be 20 cents per kWh, how much does it cost to bring this hot water to a boil?

6.5 A foam container with a thickness of 2.0 cm contains 600 g of water of 0 °C and 400 g of ice of 0 °C. Determine the time needed for all the ice to melt when the container is placed in the atmosphere with an ambient temperature of 30 °C. The external dimension of the container is 20 cm × 25 cm × 40 cm, the latent heat of melting of ice is 335 kJ/kg, and the overall heat transfer coefficient referenced to the outer surface area of the container is 5.0 W/(m^2 · K).

6.6 A heat flux meter mounted on the inner wall of the door 3-cm thick of a refrigerator reads 58 W/m^2. The temperatures of the inner and outer walls of the door are 5 °C and 24 °C, respectively. What is the thermal conductivity of the door?

6.7 A liquid food is heated by means of a double-pipe heat exchanger using hot water as the heat source. The food is flowing through the inner pipe at the rate of 2 kg/s and heated from 10 to 60 °C. Meanwhile, the hot water experiences a temperature drop from 90 to 70 °C while flowing countercurrent to the liquid food through the annular space. Let the overall heat transfer coefficient of the heat exchanger be 200 W/(m^2 · K); determine the flow rate of hot water and the required heat transfer area. Also, determine the flow rate of hot water and the required heat transfer area in the case where the hot water and the liquid food are flowed cocurrent to each other. Compare these values with those determined earlier for the case of countercurrent flow. Let the respective specific heats of the liquid food and water be 4.30 and 4.19 kJ/(kg · K).

6.8 One thousand kilograms of broccoli is precooled with cold water down to 5 °C and then loaded into a refrigerated container with a storage temperature of 2 °C before being transported to a destination located 20 h away from the loading point. In the first 1.5 h, the broccoli loaded into the container is cooled from 5 to 2 °C. The total surface area of the container walls is 30 m², and the overall heat transfer coefficient between the external atmosphere and the walls is 0.3 W/(m² · K). The specific heat of broccoli is 4.02 kJ/(kg · K) and the respiratory heat of broccoli is 35 kW/kg. If the external ambient temperature is 30 °C, what is the total quantity of heat removed by the refrigerator of the container throughout the entire transportation process, and what is the maximum rate for the heat removal?

6.9 When an asparagus of 1 cm in diameter with an initial temperature of 30 °C is cooled in an atmosphere of 3 °C, how much time is required for its center to reach 8 °C? The thermal conductivity of the asparagus is 0.604 W/(m · K), the density is 1040 kg/m³, and the specific heat is 3.31 kJ/(kg · K). In addition, consider the asparagus as an infinite cylinder, and assume that the heat transfer coefficient between it and the surrounding atmosphere is 20 W/(m² · K).

6.10 Detrimental quality changes in mandarin oranges resulted from them getting frozen on trees by cold air on winter nights pose a problem for the growers. When a mandarin orange of 11 cm in diameter with an initial temperature of 20 °C is stored in cold air of −2 °C, what will the center temperature be after 6 h? Assume that the fruit does not get frozen, and let its thermal conductivity be 0.55 W/(m · K), its density be 1030 kg/m³, its specific heat be 3.77 kJ/(kg · K), and the heat transfer coefficient between it and the surrounding atmosphere be 10 W/(m² · K).

6.11 Peas of 7 mm in diameter are spread in a layer on top of a stainless steel perforated plate, and cold air of −30 °C is blown from below the plate to suspend the peas in the air while freezing them (referred to as fluidized-bed freezing method). Assume the initial temperature of the peas is 15 °C, the wet basis moisture content is 80 % (w/w), and the freezing point is −1.0 °C, determine the length of time needed to completely freeze the peas. In addition, how much time is further required for cooling the completely frozen peas to an average product temperature of −20 °C using the same freezing equipment? Let the heat transfer coefficient between the peas and the cold air be 150 W/(m² · K), the thermal conductivity of the frozen phase in the peas be 0.5 W/(m · K), the respective specific heats of the unfrozen and frozen phases in the peas be 3.31 kJ/(kg · K) and 1.76 kJ/(kg · K), the density be 1050 kg/m³, and the latent heat of freezing of water be 334 kJ/kg.

6.12 Meatballs of 10 °C measuring 15 mm in diameter with an initial wet basis moisture content of 60 % (w/w) are placed on the conveyor belt of a tunnel freezer (Fig. 6.18b). The inlet temperature of the cooling air blown from the top through nozzles onto the meatballs is −40 °C. The air is then discharged from the freezer at −30 °C. The freezing point of the meatballs is −2 °C, at which all the water contained therein is frozen. The meatballs are further cooled after being frozen and they come out of the freezer at −20 °C. Now, find the answers to the questions below:

(a) Determine the amount of heat energy removed from 1 kg of unprocessed meatballs through the entire process of refrigerating them from the initial temperature, freezing all the water present in them at -2 °C and subsequently further cooling them down to -20 °C. Assume the specific heats of the unfrozen and frozen meatballs are 2.85 kJ/(kg·K) and 1.72 kJ/(kg·K), respectively, and the latent heat of freezing of water is 334 kJ/kg.

(b) Let the heat transfer coefficient between air and the meatballs be 15 W/(m²·K), the thermal conductivity of the frozen meatballs be 1.17 W/(m·K), and the density be 1000 kg/m³. Determine using Plank's equation the time required for refrigerating the meatballs from the initial temperature to -2 °C and then freezing them completely at the same temperature.

(c) As refrigeration further progresses, the temperature of the meatballs completely frozen at -2 °C decreases further to an average value of -20 °C. How long does this cooling process take to complete? Assume the specific heat of the frozen meatballs is 1.34 kJ/(kg·K) and the density is 980 kg/m³.

(d) Let the throughput of the freezer for the meatballs be 500 kg/h, and determine the minimum flow rate of cold air of -40 °C. Assume the specific heat of air is 1 kJ/(kg·K).

(e) Cooling of air is performed by heat exchange between air of 30 °C and a refrigerant of -60 °C through a heat exchanger to obtain the cold air of -40 °C. Assume the overall heat transfer coefficient of the heat exchanger is 100 W/(m²·K); determine the required heat transfer area. Also, assume that the refrigerant temperature stays constant throughout the heat exchange process and the refrigerant and air are flowing counter-current to each other.

6.13 A 15 % (w/w) sugar solution and a salt solution of the same concentration have water vapor pressure of 3.138 kPa and 2.825 kPa, respectively, at 25 °C. What are the respective water activity values of the solutions? The vapor pressure of pure water at 25 °C is 3.167 kPa.

Appendix

A1 Expressions of Concentration

There are various definitions for concentration of a solution. *Molar concentration* (molarity) [mol-solute/m^3-solution] is defined as the amount of substance of solute per unit volume of a solution and is often represented by the uppercase letter C. Besides, the amount of substance of solute per unit mass of the solvent of a solution is called the *molal concentration* (molality), m [mol-solute/kg-solvent]. Note that these two are different just by one letter. The weight ratio of the solute in a solution to the entire solution (sum of weights of the solute and solvent) is referred to as the *weight fraction*, w; meanwhile the molar ratio of the solute in a solution to the entire solution (sum of amounts of substances of the solute and solvent) is known as the *mole fraction*, x. The weight fraction and mole fraction when multiplied by 100 become weight percentage and mole percentage, respectively. Furthermore, the weight fraction when multiplied by 10^6 can be presented in the units of *parts per million* [ppm]. Given that the density of water approximates 1.0 g/mL, the quantity of solute in milligrams [mg] in 1 L of its very dilute aqueous solution is usually expressed in [ppm].

Concentration of alcohol is often expressed on volume basis. The volume ratio of the solute of a solution to the solution (sum of volumes of the solute and solvent) is called the volume fraction which becomes volume percentage [% (v/v)] when expressed with a denominator of 100. The mass and amount of substance of the solute in a solution per unit volume of the solution are known, respectively, as mass density [kg/m^3] and amount of substance concentration [mol/m^3]. The weight [g] of a solute present in every 100 mL of its solution is presented in the units of [% (w/v)]. When this particular value is multiplied by 10 and subsequently divided by the molar mass [g/mol] of the solute, we obtain the molar concentration [mol/L] of the solution.

© Springer Science+Business Media Singapore 2016
T.L. Neoh et al., *Introduction to Food Manufacturing Engineering*,
DOI 10.1007/978-981-10-0442-1

In the gaseous state, the concentration of a constituent substance is usually expressed in mole fraction by the lowercase letter y. In addition, the concentration may also be expressed in terms of partial pressure since the partial pressure of a particular constituent substance is proportional to its mole fraction in its gaseous mixture.

A2 Graphical and Numerical Calculus

The instantaneous speed of and distance traveled by a car is displayed, respectively, on the speedometer and odometer. The distance traveled is obtained by integrating the speed (integration). The knowledge of basic differentiation and integration is indispensable for dealing quantitatively with various phenomena in food processing. Although differentiation and integration of equations may in many cases be carried out analytically, there are also times that they may not. In addition, data may be obtained as data sets like (x_1, f_1), (x_2, f_2), ..., (x_n, f_n) instead of equations. In such cases, integration and differentiation may be performed numerically or graphically.

A2.1 Graphical Integration

As shown in Fig. A.1, integration refers to the determination of the area bounded by the integrand, $f(x)$ (quantity being integrated), and the x-axis. Hence, the value of integral can be obtained as described below.

Plot $f(x)$ on a homogeneous graph paper and connect the data points with a smooth curve. Then, cut out the area bounded by the graph and the x-axis within the intended limits of integration (integration range), and weigh it accurately on a sensitive analytical balance. Meanwhile, cut out a square or rectangular figure (of which the area is known) as the reference from the same graph paper, and measure the weight. Determine the value of integral from the weight ratio of the two figures. The above-described method is called the *graphical integration method*.

Fig. A.1 Definite integral of function $f(x)$

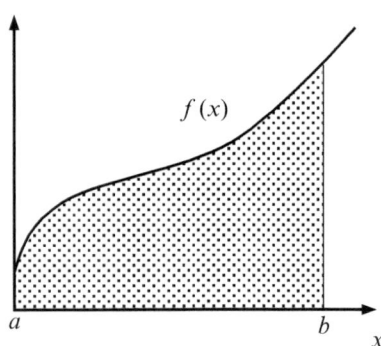

Fig. A.2 Map of the island
of Borneo

300 km

Example A.1 Determine the area of the island of Borneo as pictured in Fig. A.2 based on the same concept as the graphical integration method.

Solution By enlarging Fig. A.2 to A4 size, cutting out the figures of the island of Borneo and the 300 km × 300 km square (9×10^4 km²), and weighing the figures on a balance, the respective weights of the figures are determined to be 0.835 and 0.102 g. Hence, the area of the island of Borneo is estimated to be (9×10^4) $(0.835)/(0.102) = 7.37 \times 10^5$ km² which is very close to the literature value of 7.43×10^5 km² (with an error of less than 1 %) stated in the book of maps. ◢

A2.2   Numerical Integration

Numerical integration refers to computing the numerical value of a definite integral

$$I = \int_a^b f(x)dx \tag{A.1}$$

using numerical techniques and geometrically, as discussed in Sect. A2.1, means determination of the area bounded by the graph of $f(x)$ and the x-axis within the integration range between $x = a$ and $x = b$ (Fig. A.1). The integrand, $f(x)$, may be given by an equation or presented as tabulated data.

There exist many different methods of numerical integration such as:

1. Integration by *backward difference method*: the integration range [a, b] is divided into n equally spaced subintervals of length h, the representative value of the function f for the subinterval between x_{i-1} and x_i is approximated by $f(x_{i-1})$ (the value when $x = x_{i-1}$), and the area under the graph of the function f and between the intervals [a, b] is approximated by the sum of the rectangular areas given by $h \cdot f(x_{i-1})$ that form an area similar to the one being measured (Fig. A.3a).
2. Integration by *forward difference method*: the representative value of the function f for the subinterval between x_{i-1} and x_i is approximated by $f(x_i)$ (the value when

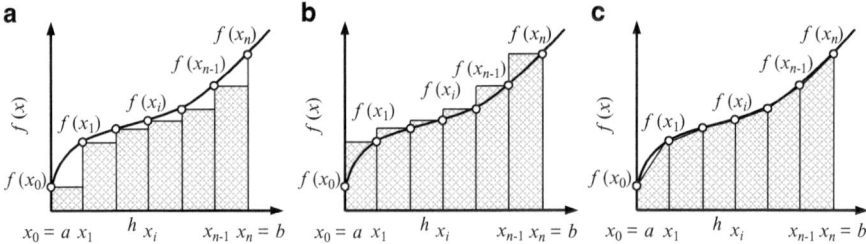

Fig. A.3 Numerical integration by (**a**) the backward difference method, (**b**) the forward difference method, and (**c**) the trapezoidal rule

$x = x_i$), and the area under the graph between the intervals $[a, b]$ is approximated by the sum of rectangular areas given by $h \cdot f(x_i)$ (Fig. A.3b).

3. Integration by *trapezoidal rule*: the representative value of the function f for the subinterval between x_{i-1} and x_i is approximated by the average of $f(x_{i-1})$ and $f(x_i)$, and the area under the graph between the intervals $[a, b]$ is approximated by the sum of the trapezoidal areas under the straight lines connecting the left and right end points of the subintervals, which are given by $h[(f(x_{i-1}) + f(x_i))/2]$ (Fig. A.3c).

The approximation obtained by the trapezoidal rule of a function is the same as the average of the approximations obtained by the backward difference and forward difference methods of that particular function. Thus, the trapezoidal rule is used to briefly find a numerical approximation for a definite integral with a certain level of accuracy.

In addition to the previously discussed methods, *Simpson's rule* is another method for numerical integration with comparatively higher accuracy. The rule fits a quadratic function (parabola) to each group of three consecutive points (for instance, in groups of (x_0, x_1, x_2), (x_2, x_3, x_4), and so on as shown in Fig. A.3). Then, the value of integral for each group is analytically computed, and finally the definite integral for the integration range between $x = a$ and $x = b$ is approximated to the sum of the integrals of all the groups. By dividing the integration range between $x = a$ and $x = b$ into an even number $(2n)$ of equal subintervals of width h where $h = (b - a)/2n$ and approximating the three data points on each consecutive pair of subintervals $x_{i-1} \leq x_i \leq x_{i+1}$ $(= x_{i-1} + 2h; i = 1, 3, 5, \ldots, 2n - 3, 2n - 1)$ with a function f, the definite integral for the integration range between $x = a$ and $x = b$ is thus given by

$$\int_a^b f(x)dx = \int_{x_0}^{x_{2n}} f(x)dx = \int_{x_0}^{x_2} f(x)dx + \int_{x_2}^{x_4} f(x)dx + \cdots + \int_{x_{2n-2}}^{x_{2n}} f(x)dx$$

$$= \frac{h}{3}(f_0 + 4f_1 + 2f_2 + 4f_3 + 2f_4 + \cdots + 2f_{2n-2} + 4f_{2n-1} + f_{2n}) \qquad (A.2)$$

Table A.1 Values of integrand with x increases by the increment of 0.1

x	0	0.1	0.2	0.3	0.4	0.5	0.6	0.7	0.8	0.9	1.0
$x^2 + 1$	1.00	1.01	1.04	1.09	1.16	1.25	1.36	1.49	1.64	1.81	2.00

Example A.2 Divide the integration range of $I = \int_0^1 \left(x^2 + 1 \right) dx$ into ten equal subintervals, and then determine numerically the definite integral of the function by the backward difference method, the forward difference method, the trapezoidal rule, and Simpson's rule. Compare the approximations with the value of integral determined analytically.

Solution Dividing the integration range $[0, 1]$ into ten subintervals and calculating the values of $x^2 + 1$ with x increases by the increment of 0.1, we obtain the values as shown in Table A.1. The value of integral determined by the backward difference method is

$$I = (0.1) \ (1.00 + 1.01 + 1.04 + 1.09 + 1.16 + 1.25 + 1.36 + 1.49$$
$$+ 1.64 + 1.81) = 1.285$$

And that given by the forward difference method is

$$I = (0.1) \ (1.01 + 1.04 + 1.09 + 1.16 + 1.25 + 1.36 + 1.49 + 1.64$$
$$+ 1.81 + 2.00) = 1.385$$

The value of integral computed by the trapezoidal rule is equal to the average of the values obtained by the backward difference and forward difference methods:

$$I = (1.285 + 1.385) \, / 2 = 1.335$$

And finally, the application of Simpson's rule gives a value of integral of

$$I = (0.1/3) \ (1.00 + (4) \ (1.01 + 1.09 + 1.25 + 1.49 + 1.81)$$
$$+ (2) \ (1.04 + 1.16 + 1.36 + 1.64) + 2.00) = 1.333$$

Meanwhile, the analytical solution of the integral equation gives

$$I = \int_0^1 \left(x^2 + 1 \right) dx = \left[\frac{1}{3} x^3 + x \right]_0^1 = \frac{1}{3} \left(1^3 - 0^3 \right) + (1 - 0) = 1.333$$

The value obtained by numerical integration by the trapezoidal rule approximates that determined by the analytical solution, indicating high accuracy of the trapezoidal rule. ◢

A2.3 Graphical Differentiation

Since differentiation (derivative) of a function means the determination of the slope of the tangent line to the graph of the function at a chosen point, we can just viscerally draw a tangent line using a ruler at the point of which we want to find the derivative (slope) and then determine the slope. However, there is another method for determining the derivative value with an improved degree of accuracy by the use of a mirror.

First, we plot the function $f(x)$ on a graph paper and connect the data point with a smooth curve. Then we hold a mirror perpendicular to the graph paper at the point, the derivative of which we are interested in. While looking from above the mirror, we rotate it until the graph and its reflection continue one another smoothly. In this position, the edge of the mirror gives the line normal to the curve at the particular point (Fig. A.4). We draw a line along the edge of the mirror and find the slope. Since the slopes of the tangent and normal lines obey this relationship, (slope of tangent) $= -1/($slope of normal$)$, we can determine the derivative value at the point from the slope of the normal line. The afore-described method is called the graphical differentiation method of which the error is relatively small if performed carefully.

A2.4 Numerical Differentiation

The method of fitting a quadratic function to three values of a function f ((x_0, f_0), (x_1, f_1), and (x_2, f_2)) from a set of tabulated data and then differentiating the fitted function analytically to estimate the derivative (differential coefficient) of the function is known as *numerical differentiation*. The values of derivative at $x = x_0$, $x = x_1$, and $x = x_2$ are given by Eq. (A.3), Eq. (A.4), and Eq. (A.5), respectively:

Fig. A.4 Graphical
differentiation using a mirror

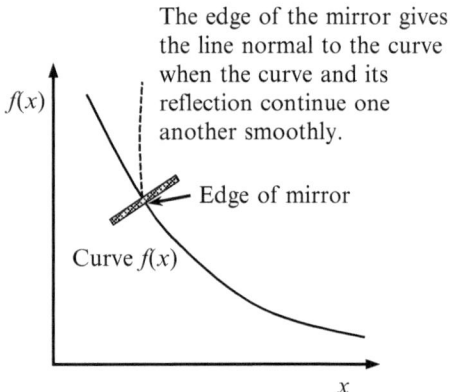

The edge of the mirror gives
the line normal to the curve
when the curve and its
reflection continue one
another smoothly.

Table A.2 Drive time and mileage

Drive time [h]	0	0.25	0.5	0.75	1.0	1.25	1.5	1.75	2.0
Mileage [km]	0	22	40	61	86	110	132	146	164

Fig. A.5 Relationship between drive time and mileage

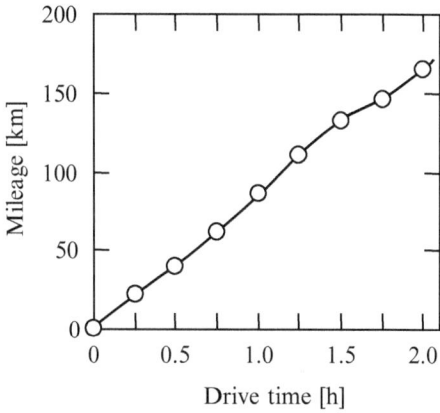

$$f'(x_0) = \frac{-3f_0 + 4f_1 - f_2}{2h} \tag{A.3}$$

$$f'(x_1) = \frac{-f_0 + f_2}{2h} \tag{A.4}$$

$$f'(x_2) = \frac{f_0 - 4f_1 + 3f_2}{2h} \tag{A.5}$$

If there exist more than three points, Eq. (A.3) and Eq. (A.5) are applicable to the determination of the differential values of the leftmost and rightmost points, respectively. Equation (A.4) is applicable to the rest of the points.

Example A.3 The backseat passenger of a car recorded every 15 min of the mileage along the travel route on an expressway (Table A.2). Determine the travel speed right after (time $t = 0$), 1 h after, and 2 h after the passenger started recording the mileage.

Solution Figure A.5 shows the graphical representation of the data set in Table A.2. First we connect the data points with a smooth curve. Then we draw the normal lines to the curve at points $t = 0$ h, 1 h, and 2 h by graphical differentiation method. Next we draw a line perpendicular to the normal line at the point and determine the slope which corresponds to the travel speed at the point. The travel speed at $t = 0$ h, 1 h, and 2 h are determined, respectively, to be 82, 100, and 85 km/h. Note that the differential values obtained by graphical differentiation may differ from person to person.

Next, we use Eq. (A.3), Eq. (A.4), and Eq. (A.5) to determine the travel speed right after (time $t = 0$), 1 h after, and 2 h after the passenger started to record the mileage, respectively. The respective travel speeds are denoted by v_0, v_1, and v_2.

$$v_0 = \frac{(-3)\,(0) + (4)(22) - (40)}{(2)(0.25)} = 96 \text{ km/h}$$

$$v_1 = \frac{(-61) + (110)}{(2)(0.25)} = 98 \text{ km/h}$$

$$v_2 = \frac{(132) - (4)(146) + (3)(164)}{(2)(0.25)} = 80 \text{ km/h} \quad \blacktriangleleft$$

A3   Variables Separable Differential Equations

The equations that can be transposed to the form below

$$g(y)dy = f(x)dx \tag{A.6}$$

by algebraic manipulation in the first-order differential equations (rearrangement of equations) are known as the *variables separable differential equations* because the variable x appears only on the right side and the variable y appears only on the left side of the equations. Integrating the both sides of Eq. (A.6), we obtain:

$$\int g(y)dy = \int f(x)dx \tag{A.7}$$

A *generalized solution* for Eq. (A.6) can be obtained if f and g can be integrated. Nonetheless, in the fields of agriculture and engineering, particular solutions that satisfy certain conditions are often required rather than generalized solutions. In these cases, it is necessary to determine the integration constant, c, by using the *initial condition* (or the *boundary condition*).

Example A.4 The decrease rate of the number of viable microorganism cells, N, in a sterilization operation is proportional to the number of viable microorganism cells, N. That is, the kinetics of a sterilization process is expressed by a first-order reaction equation:

$$-\frac{dN}{dt} = k_d N \tag{A.8}$$

where k_d denotes the death rate constant. Determine the relationship between the sterilization time and the number of viable microorganism cells. Let the initial number of viable microorganism cells be denoted by N_0.

Solution Variables in Eq. (A.8) are separable as shown by the following equation in which the left side of the equation consists only of the variable N, while the right side consists only of the variable t.

$$-\frac{dN}{N} = k_d dt \tag{A.9}$$

Integrating the both sides of Eq. (A.9) gives

$$\int \frac{dN}{N} = -k_d \int dt + c \tag{A.10}$$

$$\ln N = -k_d t + c \tag{A.11}$$

Here, by using the initial condition of $N = N_0$ when $t = 0$, c is determined to be

$$c = \ln N_0 \tag{A.12}$$

Hence, the solution for Eq. (A.8) is given by

$$\ln N = -k_d t + \ln N_0 \tag{A.13}$$

Eq. (A.13) can be rewritten as follows:

$$N = N_0 e^{-k_d t} = N_0 \exp\left(-k_d t\right) \tag{A.14}$$

In the above-described steps, we used the initial condition to determine c after obtaining the generalized solution. However, the calculation process can be simplified by integrating the both sides of Eq. (A.9) using the conditions of $N = N_0$ when $t = 0$ and $N = N$ when $t = t$:

$$\int_{N_0}^{N} \frac{dN}{N} = -k_d \int_{0}^{t} dt \tag{A.15}$$

The following equation is obtained by definite integration of Eq. (A.15):

$$\ln \frac{N}{N_0} = -k_d t \tag{A.16}$$

Rewriting Eq. (A.16) gives

$$N = N_0 e^{-k_d t} \tag{A.17}$$

Here, we managed to arrive at the same solution skipping a few steps of calculation.
◢

A4   Estimation of Parameters with Microsoft Excel®

If the variables x and y obtained from an experiment or observation are related by

$$y = ax + b \qquad\qquad (A.18)$$

then the plot of y against x will yield a straight line. The approach of plotting the actual measurement values into a graph and estimating the parameters a and b in Eq. (A.18) from the slope and y-intercept of the straight line, respectively, is called *parameter estimation*. Imagine that we obtain the symbols in Fig. A.6 by plotting the actual measured data points of y at their respective x values. Now, we have to decide on how to draw a straight line connecting the data points in order to find the constants a and b. As the straight line may be drawn differently from person to person (compare the solid and dashed lines), not surprisingly, the estimated values of a and b will differ as well. Even though it may not be such a big problem to just viscerally draw a seemingly best straight line that passes through all the data points, a reasonable method by which the same values for a and b can be estimated regardless of by whom the estimation is performed is still desirable. The *method of least squares* is one of the most widely used approaches for this purpose. Let the y value calculated by Eq. (A.18) for an x value and the actual measurement value of y at the corresponding x value be y_{cal} and y_{obs}, respectively. $y_{obs} - y_{cal}$ (or $y_{cal} - y_{obs}$) is referred to as a *residual* (Fig. A.7). It may appear to be good to estimate the constants a and b just by minimizing the residuals of all measured data points, but since the residuals for the data points may take either positive or negative values, the sum of the residuals may somehow approach 0 as they cancel out each other. Since squared numbers are positive numbers, we first square the residuals of the data points to make sure they are always positive and then find the values of a and b that minimizes the *sum of the squares of the residuals* (sum of squared residuals, M). Such a procedure is known as the method of least squares. In particular, when the relationship between variables x and y is described by a linear function as in

Fig. A.6 Drawing of straight lines that visually best fit the actual measured data points

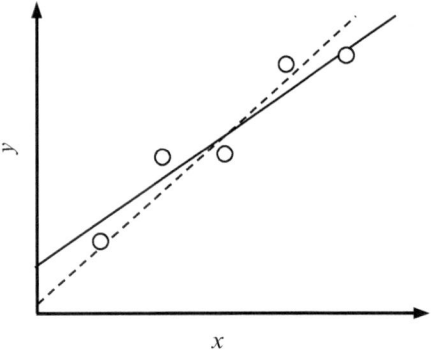

Fig. A.7 Basics for the method of least squares

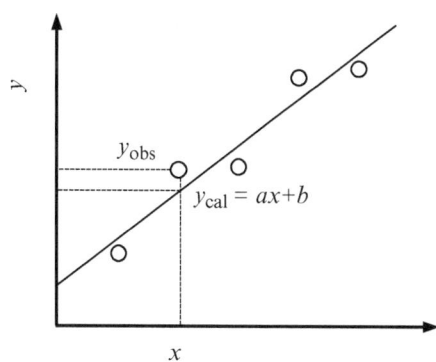

Table A.3 Determination of bulk density of a wheat flour

Volume, V [cm^3]	46	88	133	185	228
Weight, W [g]	237	255	286	309	338

Eq. (A.18), the method is called the *linear least squares* as opposed to the *nonlinear least squares* for estimating the parameters in a relationship between variables x and y which is not expressed by a straight line (but a curve).

As one of the most extensively used spreadsheet software, Microsoft Excel® is capable of performing the function of estimation of parameters that best fit a data set (plotted on a graph) when the relationship between the two variables x and y is described by, for instance, a first-order equation (linear approximation) as in Eq. (A.18) or other standard function forms such as the logarithmic, exponential, and power functions.

Example A.5 The volume, V, of a wheat flour was measured by gently adding 3 tablespoons of the flour into a 250-mL graduated cylinder. And the weight, W, of the graduated cylinder was measured on a scale. These steps of operation are repeated and the measurements are obtained as summarized in Table A.3. Determine the bulk density, ρ, of the wheat flour.

Solution The wheat flour in the graduated cylinder contains space in between the particles, and the bulk density of the flour is, according to the definition, the weight per unit volume of the flour, which includes this interparticle space. The volume, V, and the weight, W, obtained from the measurement mentioned above obey the relationship described by the following equation, and the bulk density, ρ, can thus be determined as the slope of the straight line yielded by the plot of W versus V.

$$W = \rho V + W_0 \tag{A.19}$$

where W_0 represents the weight of the empty cylinder. We will determine the bulk density of the flour and the weight of the empty cylinder by the use of the parameter

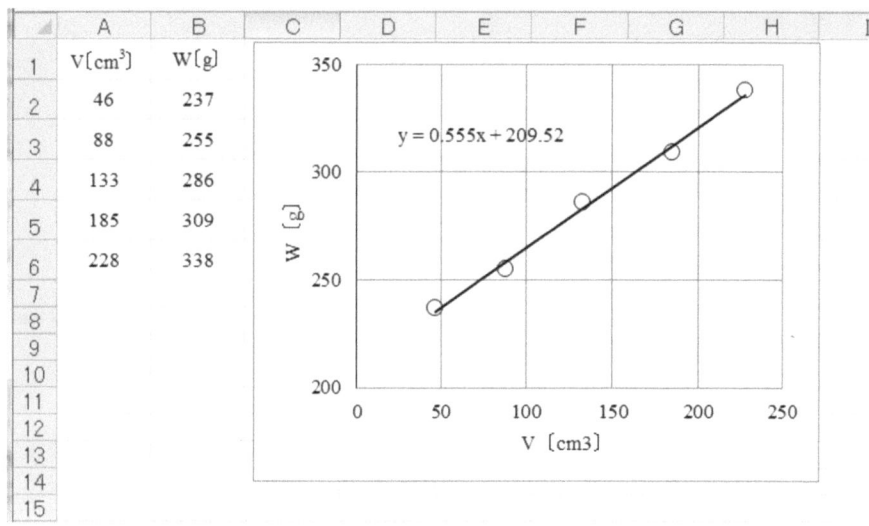

Fig. A.8 Estimation of bulk density of a wheat flour

estimation function of Microsoft Excel®. First, we enter the data set in Table A.3 on a worksheet. Next, we select the cells from A2 to B6 and then create a graph by clicking the [Insert] tab and selecting [Chart] → [Scatter] → [Scatter with only markers]. We may also make some modifications to the settings in [Chart Tools] in order to improve the clarity of the graph. Subsequently, we left-click on one of the markers and the [Chart Tools] toggle is activated now. Here we select the [Layout] tab and click the [Trendline] button to bring up the menu and then select [More Trendline Options] to activate the Format Trendline dialog box. In Trendline Options of the dialog box (which can also be reached by right-clicking on one of the markers and selecting Add Trendline from the pop-up menu), select the Linear radio button and the straight line that best fits the data will appear on the graph. In addition, if we tick the check box of Display Equation on chart in Trendline Options, the equation of the straight line ($y = 0.555x + 209.52$) will be displayed on the graph (Fig. A.8). Since the slope of the straight line gives the bulk density, ρ, hence ρ is determined to be 0.555 g/cm^3. Furthermore, W_0 is obtained from the y-intercept as 209.52. Also, if we tick the check box of Display R-squared value on chart in Trendline Options, the R-squared value will be shown on the graph (in this example, $R^2 = 0.9945$). R^2, also known as the coefficient of determination, is an indicator of how well the estimated model fits the actual data, and it can take any value between 0 and 1. The closer the R^2 value is to 1, the better fit the trendline is to the data. ◢

Table A.4 Substrate concentration and specific proliferation rate

C_S [kg/m^3]	0.036	0.089	0.167	0.291	0.420
μ_{obs} [1/h]	0.0529	0.0942	0.1251	0.1488	0.1615

Example A.6 The specific proliferation rate, μ, of a type of microorganism was measured at various substrate concentrations, C_S, and the results are tabulated in Table A.4. Determine the maximum specific proliferation rate, μ_{max}, and the saturation constant, K_S, that best fit these actual measured values when the relationship between μ and C_S is expressed by Eq. (A.20) (known as the Monod equation):

$$\mu = \frac{\mu_{max} C_S}{K_S + C_S} \tag{A.20}$$

Solution Taking the reciprocal of both sides of Eq. (A.20), the equation is transformed into

$$\frac{1}{\mu} = \frac{K_S}{\mu_{max}} \frac{1}{C_S} + \frac{1}{\mu_{max}} \tag{A.21}$$

Now, Eq. (A.21) has the same form as Eq. (A.18), and the plot of $1/\mu$ against $1/C_S$ will yield a straight line, from the y-intercept and slope of which μ_{max} and K_S can be determined, respectively. The graph of μ versus C_S gives a curve, indicating a nonlinear relationship. This approach of transforming an equation into a linear form like Eq. (A.21) is called the linearization of a nonlinear equation. If a nonlinear equation can be linearized in this way, the parameters can be estimated as per the method described in Example A.5.

As shown in Table A.4, even when the intervals in C_S are nearly equal, the reciprocal of C_S ($1/C_S$) has unequal intervals, resulting in the gaps between data points in the plot of $1/\mu$ against $1/C_S$ swelling in the positive direction of the horizontal axis. In such a case, if we try to draw a visually best-fitting straight line, we tend to draw a line that passes through the data point of the lowest substrate concentration (the rightmost data point on the graph) at which the specific proliferation rate is hard to determine experimentally at a satisfactory level of accuracy. Furthermore, even by the method of linear least squares explained in Example A.5, this particular data point will be equally weighted as other data points. By multiplying the both sides of Eq. (A.21) by C_S and swapping positions between the first and second terms on the right side, we obtain the following equation:

$$\frac{C_S}{\mu} = \frac{1}{\mu_{max}} C_S + \frac{K_S}{\mu_{max}} \tag{A.22}$$

Eq. (A.22), as well, has the same form as Eq. (A.18), and the plot of C_S/μ against C_S will yield a straight line of slope $1/\mu_{max}$ and y-intercept K_S/μ_{max}. μ_{max} can be determined from the slope and then K_S from μ_{max} and the y-intercept. In this case, C_S (instead of $1/C_S$) is plotted on the horizontal axis and therefore the data points have nearly even intervals.

We mentioned previously that the method of least squares estimates the parameters of a function that minimizes the sum of the squares of the differences (residuals) between the actual measured and calculated values. The same concept also applies to the case of a nonlinear equation like Eq. (A.20), which means to estimate the values for μ_{max} and K_S that minimize the sum of the squares of the residuals between the actual measured values of the specific proliferation rate, μ_{obs}, and the values calculated by Eq. (A.20), μ_{cal}:

$$M = \sum (\mu_{obs} - \mu_{cal})^2 \tag{A.23}$$

Microsoft Excel® has a very useful tool called the *Solver Add-in* that can be utilized for this estimation purpose.

First, let us get the worksheet ready. The estimates of μ_{max} and K_S will be shown in cells B4 and B5, respectively. We enter the data on a worksheet (Fig. A.9): the values of substrate concentration in cells from C2 to G2 and the actual measured values of the specific proliferation rate, μ_{obs}, in cells from C3 to G3. Then, we determine the calculated values of the specific proliferation rate, μ_{cal}, by using Eq. (A.20). However, at this point the estimates of μ_{max} and K_S are still yet to be obtained. For the application of the Solver Add-in for parameter estimation, we are required to assign appropriate initial values for the parameters to be estimated before the values can be sequentially improved until the optimal values are obtained (this process is run automatically by the Solver Add-in). Therefore, we will need to roughly estimate the initial values of μ_{max} and K_S. From the data points in the graph, it seems that the value of μ approaches asymptotically $\mu = 0.18$ h^{-1} for substrate concentrations, C_S greater than 0.4 kg/m^3. Thus we assume the maximum specific proliferation rate, μ_{max}, to be 0.18 h^{-1}. In addition, because K_S is equal to the substrate concentration at the μ value half of μ_{max} ($\mu = \mu_{max}/2$), its initial value is estimated from the graph to be around 0.08 kg/m^3. Next, we enter these values as the initial values for μ_{max} and K_S in cells B4 and B5, respectively.

After entering the initial values of μ_{max} and K_S, we use the calculation function in Microsoft Excel® to compute μ_{cal} (Fig. A.9). In the calculation of μ_{cal}, the references in the calculation equation that refer to the cells in which the values of μ_{max} and K_S are held remain fixed even though C_S changes, whereas the reference that refers to the cell where the C_S value is held changes. Therefore, we use absolute references in the calculation equation to refer to the cells where the values of μ_{max} and K_S are held and a relative reference to refer to the cell where the C_S value is held. For example, in the calculation equation "=$B$4*C2/($B$5+C2)" of cell C6, the

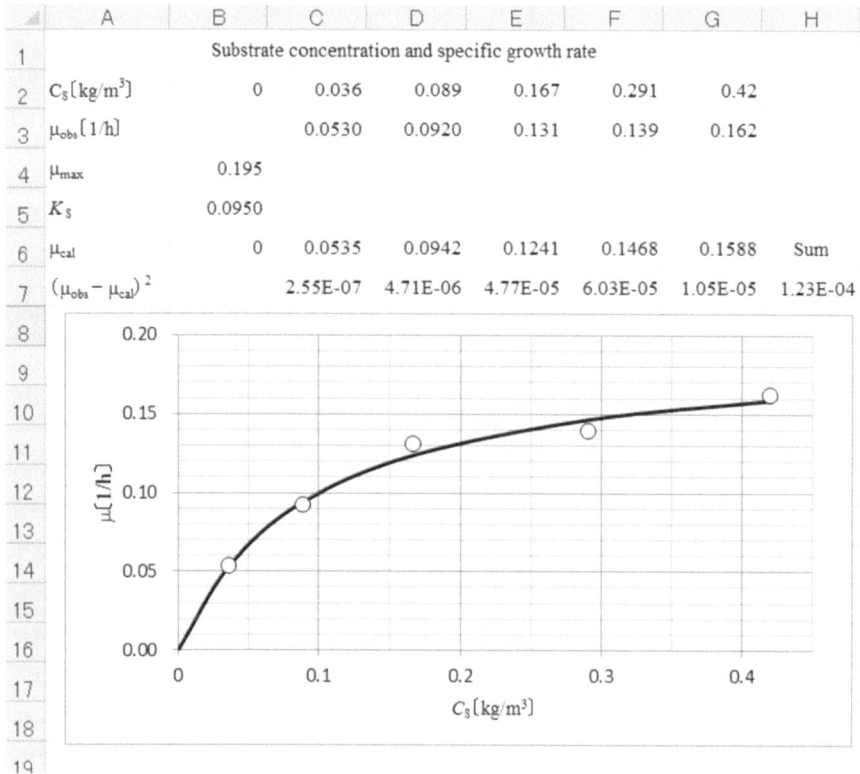

	A	B	C	D	E	F	G	H
1			Substrate concentration and specific growth rate					
2	C_S[kg/m³]	0	0.036	0.089	0.167	0.291	0.42	
3	μ_{obs}[1/h]		0.0530	0.0920	0.131	0.139	0.162	
4	μ_{max}	0.195						
5	K_S	0.0950						
6	μ_{cal}	0	0.0535	0.0942	0.1241	0.1468	0.1588	Sum
7	$(\mu_{obs}-\mu_{cal})^2$		2.55E-07	4.71E-06	4.77E-05	6.03E-05	1.05E-05	1.23E-04

Fig. A.9 Estimation of the parameters of the Monod equation

cell references with the dollar symbols ($B$4 and $B$5) are absolute references, whereas the one without (C2) is a relative reference. Hence, if we copy cell C6 and paste it in cell D6 to calculate the value of μ_{cal} at $C_S = 0.089$, $B$4 and $B$5 remain the same in the calculation equation in cell D6, while the term C2 adjusts automatically to D2. Likewise, we determine the values of μ_{cal} by copying cell C6 to the cells from E6 to G6. Next, we write another calculation equation "=(C3–C6)^2" in cell C7 to determine the squared residual at $C_S = 0.036$. We copy cell C7 to the four cells to the right on the same row to determine the squared residual at each substrate concentration. And after that, we write by using the SUM function the calculation equation "=SUM(C7:G7)" in cell H7 to calculate the sum of squared residuals, M.

Now, we use the Solver Add-in to estimate μ_{max} and K_S that minimize the value in cell H7 (sum of squared residuals, M). First, we bring up the Solver Parameters dialog box by left-clicking the [Data] tab and selecting the [Solver] button. Because we want to estimate the values in cells B4 and B5 that minimize the value in cell

H7, we input the absolute cell reference $H$7 in the Set Objective box by clicking directly cell H7 on the worksheet. Next, we click the Min radio button because we want the value of the objective cell to be the smallest possible. Subsequently, we enter the absolute references of cells B4 and B5 ($B$4:$B$5) in the By Changing Variable Cells box or click the cells directly on the worksheet to find the values for μ_{max} and K_S. For this example, we will leave the Subject to the Constraints box blank. If there are any constraints to be applied such as equality and inequality, they may be entered in this box.

Now, by just a click on the [Solve] button, the calculation process runs immediately, and later the Solver Results dialog box pops up. Note that the values in cells B4, B5, and also H7 are now different from the initial values. Here, the values displayed in cells B4 and B5 are the respective values of μ_{max} and K_S that we wanted to estimate: $\mu_{max} = 0.195$ h^{-1} and $K_S = 0.095$ kg/m^3. The zeros entered in cells B2 and B6 are just for the sake of passing the curve of μ_{cal} through the origin. ◢

The GAB equation (A.24) which describes the moisture sorption isotherm of durum semolina covered in Example 3.13 is a nonlinear function. The GAB equation has three parameters q_m, c, and K which can be estimated also by using the Solver Add-in.

$$q = \frac{q_m c K a_w}{(1 - K a_w)(1 + (c - 1) K a_w)} \tag{A.24}$$

Because there are three parameters in Eq. (A.24), the issue about how to estimate their initial values becomes critical. Let us consider the cases of low water activity; i.e., $a_w < 1$. Meanwhile, c is generally greater than 1, and if we assume $c \gg 1$, then we can also assume $c - 1 \approx c$. By these assumptions, Eq. (A.24) can thus be approximated as follows:

$$q = \frac{q_m c K a_w}{1 + (c - 1) K a_w} \approx \frac{q_m a_w}{1/cK + a_w} \tag{A.25}$$

Eq. (A.25) has the same form as Eq. (A.20). Thus similar to Example A.6, from the plot of q against a_w, the initial value for q_m (the maximum amount of moisture sorbed when the solid surface (surface of durum semolina) is covered with a monolayer of water molecules) is estimated from the data points at low a_w to be 0.09 g-water/g-d.m. As discussed previously in Example 3.13, the initial value of K is estimated based on past empirical knowledge to be 1. Meanwhile, c is estimated to be approximately 10 from an arbitrary pair of a_w and q values in Table 3.7 and the estimated initial values of $q_m \approx 0.09$ g-water/g-d.m. and $K \approx 1$. By using the Solver Add-in following the same procedure as described in Example A.6 and entering these estimated values as the initial values, we finally obtain $q_m = 0.0879$ g-water/g-d.m., $K = 0.672$, and $c = 14.2$.

Major Physical Constants

Avogadro constant	6.022×10^{23}
Acceleration of gravity	9.807 m/s^2
Gas constant	32.174 ft/s^2
	8.314 J/(mol·K)
Molar volume of ideal gas at 1 atm	0.08205 L·atm/(mol·K)
	1.987 cal/(mol·K)
	1.986 Btu/(lb·mol·K)
	22.41×10^{-3} m^3/mol
Average molar mass of air	359.0 ft^3/mol
	28.97 g/mol
Conversion of °C to K	$T\,[\text{K}] = t\,[°\text{C}] + 273.15$
Conversion between °C and °F	$t\,[°\text{C}] = (t'\,[°\text{F}] - 32) \times (9/5)$
	$t'\,[°\text{F}] = (9/5)t\,[°\text{C}] + 32$

Unit Conversion Factors

Length

m	ft	in
1	3.281	39.37
0.3048	1	12
0.0254	0.08333	1

Mass

kg	lb	t
1	2.205	0.001
0.4536	1	4.536×10^{-4}
1000	2205	1

Density

kg/m³	g/cm³	lb/ft³
1	0.001	0.06243
1000	1	62.43
16.02	0.01602	1

Force

N	dyn	kgf
1	10^5	0.102
10^{-5}	1	0.102×10^{-5}
9.807	9.807×10^5	1

Pressure

Pa	kgf/cm²	mmHg	atm	bar
1	1.02×10^{-5}	7.501×10^{-3}	9.869×10^{-6}	10^{-5}
9.807×10^4	1	735.6	0.9678	0.9807
133.3	1.36×10^{-3}	1	1.316×10^{-3}	1.333×10^{-3}
1.013×10^5	1.033	760	1	1.013
10^5	1.02	750.1	0.9869	1

Energy, Work, and Heat

J	erg	cal	Btu	kW·h
1	10^7	0.2389	9.478×10^{-4}	2.777×10^{-4}
10^{-7}	1	2.389×10^{-8}	9.478×10^{-11}	2.777×10^{-11}
4.186	4.186×10^7	1	3.968×10^{-3}	1.162×10^{-6}
1055	1.055×10^{10}	252	1	2.931×10^{-4}
3.6×10^6	3.6×10^{13}	8.599×10^5	3412	1

Power

W	kgf·m/s	PS	HP	cal/s
1	0.102	1.36×10^{-3}	1.341×10^{-3}	0.2389
9.807	1	0.01333	0.01315	2.343
735.5	75	1	0.9863	175.7
745.7	76.04	1.014	1	178.1
4.186	0.4269	5.692×10^{-3}	5.615×10^{-3}	1

Thermal Conductivity

W/(m·K)	kcal/(m·h·°C)	Btu/(ft·h·°F)
1	0.8598	0.5778
1.163	1	0.6723
1.731	1.488	1

Heat Transfer Coefficient

W/(m²·K)	kcal/(m²·h·°C)	Btu/(ft²·h·°F)
1	0.8598	0.1762
1.163	1	0.2048
5.678	4.882	1

Viscosity

Pa·s	kg/(m·h)	P	lb/(ft·s)
1	3600	10	0.672
2.778×10^{-4}	1	2.778×10^{-3}	1.867×10^{-4}
0.1	360	1	0.0672
1.488	5357	14.88	1

Answers to Selected Exercises

Chapter 2

2.1 (a) 8.317 m^3 · Pa/(mol · K). (b) 333.6 kJ/(kg · K). (c) 3.168 kPa. (d) 8.90 × 10^{-4} Pa · s. (e) 0.607 W/(m · K).

2.2 The wet basis moisture content = 41.6 %. The dry basis moisture content = 71.1 %.

2.3 0.978.

2.4 10.7 kg.

2.5 Water evaporated = 0.56 kg. The weight ratios of solid and water = 0.3:0.14.

2.6 Product of sugarcane solution = 5135 kg/day. Water evaporated = 4865 kg/day.

2.7 Starch granules needed = 10,556 kg/h. Water evaporated = 8081 kg/h.

2.8 26.3 °C.

2.9 Accumulation of cellulose = 8 × 10^4 kg/day. Accumulation of microorganisms = 5.5 × 10^3 kg/day.

2.10 (a) 8770 kg. (b) 8233 kg. (c) The dried soybean flakes = 7670 kg. The protein composition = 44.3 %(w/w).

2.11 The unhydrolyzed starch = 0.005. Water = 0.701. Glucose = 0.294.

2.12 Ethanol concentration = 9.73 %(w/w). Distillate = 39.5 kg/h. Bottoms = 60.5 kg/h.

2.13 The unrecoverable mayonnaise = 6.4 %.

2.14 pH = 0.699.

2.15 pH = 0.998.

2.16 ρ_S = 1.59 g/cm^3. ρ_W = 0.997 g/cm^3.

2.17 Q = 0.12 m^3/min.

2.18 a = 1634. b = 4.462 × 10^{-3}.

2.19 a = 2.037. b = 0.3186.

Chapter 3

3.2 (II) n = 1.25. d_e = 40.2 μm. (III) The number mean diameter = 1.78 μm. The surface mean diameter = 1.88 μm.

3.1 The Martin diameter = 40 μm. The Krumbein diameter = 41 μm.

3.3 The median diameter = 30.0 μm and the mode diameter = 44.4 μm.

3.4 (II) 193 kPa. (III) The breaking strain = 0.674. The breaking stress = 518 kPa. (IV) 0.729 J.

3.5 0.638.

3.6 The critical moisture content = 0.41. The equilibrium moisture content = 0.1.

Chapter 4

4.1 For a single extraction $E = 0.924$. For three-step batch extraction, $E = 0.955$.

4.2 39 kg/m^3, cheese.

4.3 Water evaporated $= 194.6$ kg/h. The amount of concentrate $= 205.4$ kg/h.

4.4 The feed rate of raw juice $= 1.36$ kg/s. The amount of saturated steam $= 0.527$ kg/s.

4.5 0.0199 kg-water/kg-dry air.

4.6 The fresh inlet air $= 95.59$ kg-dry air/h. The recycle air $= 27.3$ kg-dry air/h. The product $= 91.4$ kg/h.

Chapter 5

5.3 The critical micelle concentration $= 0.10$ mmol/L. The surface excess $= 4.92 \times 10^{-6}$ mol/m^2. The area occupied $= 3.38 \times 10^{-19}$ m^2/molecule $= 0.338$ nm^2/molecule.

5.4 $n = 0.7$. $\eta = 22.9$ Pa$\cdot$s$^{0.7}$.

5.6 $\lambda = 1.26$ s. $\mu = 1.16$ Pa$\cdot$s.

Chapter 6

6.1 $D = 6.57$ min.

6.2 $k = 0.348$ h^{-1}.

6.3 The activation energy $E_o = 61.7$ kJ/mol. The frequency factor $A_0 = 1.18 \times 10^9$ h^{-1}.

6.4 (a) 5.02 min. (b) 2.1 cents.

6.5 32.4 min.

6.6 $k = 0.0916$ W/m$\cdot$K.

6.7 The flow rate of hot water $W_h = 5.13$ kg/s. The required heat transfer area $A = 49.7$ m^2 for the countercurrent flow and $A = 63.9$ m^2 for the cocurrent flow.

6.8 The total heat removed $= 0.252$ kW. The maximum heat removal rate $= 0.035$ kW.

6.9 559 s $= 9.32$ min.

6.10 0.24 °C.

6.11 The time needed to completely freeze $= 132.9$ s. The time needed to further cool to -20 °C $= 23.4$ s.

6.12 (a) 280 kJ/kg. (8b) 20.2 min. (c) 24.0 min. (d) 140,000 kJ/h. (e) 8.36 m^2.

6.13 a_w of sugar solution $= 0.991$. a_w of NaCl solution $= 0.892$.